Ask Uncle Albert

100 ½

Tricky Science Questions Answered

RUSSELL STANNARD

illustrated by John Levers

faber and faber

First published in 1998
by Faber and Faber Limited
3 Queen Square London WC1N 3AU

Typeset by Faber and Faber Ltd
Printed in England by Mackays of Chatham plc, Chatham, Kent

A CIP record for this book
is available from the British Library

ISBN 0–571–19436–2

4 6 8 10 9 7 5

Contents

Foreword

Many young readers of my Uncle Albert books write to me (or to Uncle Albert) with their questions about anything and everything under the Sun. These letters gave rise to *Letters to Uncle Albert* and *More Letters to Uncle Albert*.

In this new book, I have selected the best of the scientific questions from those two volumes and added a host of new ones from more recent letters.

Some of my replies to the letters now include a Quick Quiz question (complete with answer at the back of the book). These will help you to find out how well you are understanding the science.

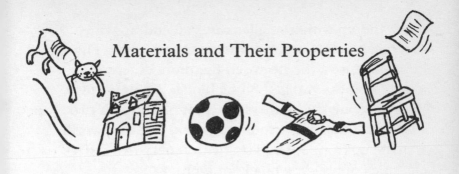

Materials and Their Properties

What is an atom made of?
Lin (aged 8)

Everything is made of atoms. Take, for example, a grapefruit. You cut it in half. Then you carry on slicing it. Imagine doing this twenty-seven times. What do you get? A slice one atom thick. Chopping up this slice, you end up with a single atom. A tiny piece of grapefruit not worth eating? No. This is not even grapefruit. (I say 'imagine' doing this because you would need a very sharp knife to do it – and they don't come that sharp – and if they did, they would be dangerous, and you shouldn't be playing with it anyway.)

Atoms are the building blocks out of which the world is made. Animals, dirt, houses, footballs, clothes, the chair you are sitting on, this page you are reading – they are all made out of atoms stuck together.

I like to think of the world as a gigantic Legoland! When you take a model Lego car to pieces you

1

don't end up with a smaller car; you end up with separate blocks – blocks with studs and holes. The more expensive the Lego set, the more different kinds of blocks you get. Some blocks are tiny – they have only one stud; others are a bit bigger with two studs, yet others have four studs, six, etc. The world is made up from a 'Lego set' with ninety-two different kinds of building block, or atom.

The word 'atom' means 'something that cannot be cut'. So, is that *it* – you can't cut an atom in half?

That's what people used to think. But now we know that isn't true. When you take an atom to pieces, you find, at the very centre, a tiny ball. It is called the *nucleus* of the atom. And surrounding the nucleus are even tinier balls called *electrons*. In fact, electrons are so tiny, they might not have any size at all. They buzz around the nucleus like bees round a hive.

Why do the electrons stay close to their nucleus? It is all due to something called *electric charge*. It comes in two kinds: positive charge carried by the nucleus, and negative charge on the electrons. Positive and negative charges pull on each other; they like to get together. It's this electric force from the nucleus that stops the electrons from wandering off.

An atom is mostly empty space. If you were to imagine it blown up to the size of an airport, the nucleus would be no bigger than a golf ball on the runway. As for the electrons, they would be smaller

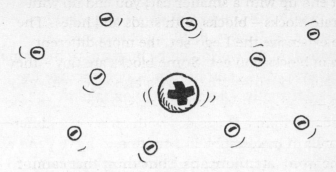

than peas, whizzing round the outside of the air-port. So, the next time your mum says you can't have a second helping of stodgy pudding, try telling her, 'But Mum, Uncle Albert says it's mostly empty space.' (You never know, it *might* work.)

What makes the ninety-two atoms different from each other? Two things. Firstly, they have a different-sized nucleus. Secondly, they have a different number of electrons. The lightest atom has only one electron (like the smallest Lego brick having only one stud); the next one up has two electrons, the next three; and so on . . . all the way up to the heaviest atom which has (you've guessed it) ninety-two electrons.

When atoms are put together, the electrons rearrange themselves around their nuclei so as to make the atoms stick. (Notice that we say 'nuclei', not 'nucleuses'. It's not my fault; I didn't invent spelling.) The electrons rearrange themselves in the same way as the studs and holes in the Lego blocks fit into each other. And just as you can build all sorts of different things out of the same set of Lego blocks – a toy tractor, a ship, a house, etc. – so everything in the world can be built out of the same ninety-two types of atom. You simply arrange the pieces differently. Clever, don't you think?!

I am interested in atoms and have read a lot about them, but have not managed to find out what the nucleus is made from.
Christopher Llewellyn

The nucleus of an atom is made from two types of particle: the *neutron* and the *proton*. These are very like each other, except that the proton carries a positive electric charge whereas the neutron has none (we say it is electrically 'neutral'; that's where its name comes from). In the answer to the last question we noted that the nucleus had positive charge; now we know that this is because of the charge on its protons.

With its neutrons and protons, the nucleus looks a bit like a raspberry.

So is that as much as we can say about the nucleus? Well, having found out that it is made of neutrons and protons, the next question is obvious: What are neutrons and protons made of? The answer is *quarks*. These are so tiny they seem to have no size at all. Each neutron and proton is made up of three quarks buzzing around each other.

So we have a tight swarm of quarks making up the nucleus, and outside that, a looser swarm of electrons to complete the atom.

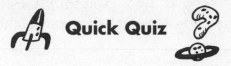 **Quick Quiz**

Do the neutrons in the nucleus help to hold the electrons in the atom?

(You will find the answers to these and other Quick Quiz questions at the back of the book.)

If you caught an electron you should be able to cut it again and again and again, etc. Is that correct?
Christopher Martin (aged 10)

Earlier I said how electrons were very, very small. In fact, we think they have no size at all – just like the quarks. And if they have no size, then you can't cut them. And if you can't cut them, then it doesn't make sense to think of them as being made up of even smaller parts; there is obviously nothing smaller than 'no size at all'.

Which I think is a good thing. After all, we have seen how everything is made of atoms, atoms are made of electrons and a nucleus, the nucleus is made of neutrons and protons, and neutrons and protons are made of quarks. If we did find quarks and electrons were made from something else, you would then want to know what *they* were made of, and so on. This could go on for ever – and we wouldn't have space for any more questions!

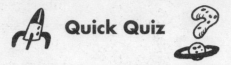 **Quick Quiz**

Why is 'atom' not a very sensible name?

 How many atoms would make up my family of four?
Sarah (aged 11)

100,000 million million million million! (That's assuming your mum and dad are not too fat.)

This is because atoms are so very, very tiny.

What does a big number like that mean? Imagine building a sand-castle. Your castle is so big, it covers not only all the beach but all of Britain - every square inch of it. And that's not all: the castle is 10,000 miles high! The number of grains of sand in your castle would be about the same as the number of atoms in your family.

Now you're asking, 'How does he know that?' Well, I didn't build a sand-castle like that – obviously. But I did count grains of sand. I got a teaspoon of sand, poured it out on to the kitchen top, spread it out, took a knife and divided it into two piles. I then spread out one of the piles, and divided that in two. I then divided one of *those* piles

6

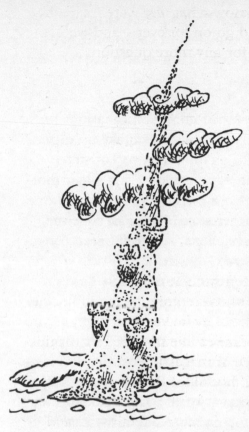

in two, etc. That carried on until I ended up with a weeny, weeny pile. I then got out a magnifying glass and counted how many grains were in that smallest pile. I multiplied that by two, then that by two, then that by two . . . until I had done it the same number of times as I had divided the sand. That told me how many grains there were in a tea-spoon of sand.

Next I worked out how many teaspoons I would need to get the same number of grains as there are atoms in your family – and that told me how big the sand-castle would be.

After that, I got into trouble for having made a mess on the kitchen top!

On holiday when I was practising my surfing in the sea, I kept falling off my board into the sea. That made me wonder why is the sea wet?
Miranda Marsh (aged 9)

Atoms stick together in different combinations to form *molecules*. Different molecules make up different substances. So, for example, a drop of water consists of lots of water molecules. Each water molecule is made up of two atoms of hydrogen and one of oxygen (if you're interested). You can't see the individual molecules because, as I told Sarah (question 4), atoms are incredibly small.

Not only do the atoms in the molecule hold on tightly to each other so as to make up the molecule, but when you get a lot of molecules together, as you do in a droplet of water, each molecule gently pulls on its neighbours. When that happens, the molecules all want to get as close as they possibly can to each other, and the best way of doing that is if they end up as a round drop (a bit like the way rugby players all try to get the same ball and end up as a round pile on the ground). The droplet of water ends up round because of the forces between the water molecules.

These molecules not only pull on each other, they can also pull on molecules belonging to any surface they come into contact with – whether that is rainwater on a window-pane, or sea water in contact with you when you fall off the surfboard. When we say 'water is wet', we simply mean that water molecules pull hard on molecules belonging to surfaces like glass or skin. They stick to surfaces. With other kinds of liquid it is not like that. Molecules of mercury – the silvery stuff you get in thermometers – pull much harder on each other than they do on molecules belonging to the walls of the glass tube. That's why they don't wet the inside of the tube.

Don't worry about falling off the board. I think anyone who manages to spend any time at all on one of those things is quite brilliant! Besides, you soon get dry again. That's because the jiggling water molecules eventually escape the pull of their neighbours, and drift off. That's when we say the water has 'evaporated'.

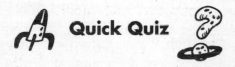 **Quick Quiz**

Why can't we see the water molecules that have evaporated?

If you need oxygen to breathe, why can't you breathe carbon dioxide because it has oxygen in it? If oxygen is O and carbon dioxide is CO_2, that's twice as much oxygen.
Christopher Parker (aged 8)

That's a very good question. In fact, I have just been telling Miranda how the water molecule is made of two atoms of hydrogen and one atom of oxygen. Hydrogen gas is very dangerous; it easily catches fire. As for oxygen, fires need oxygen to burn. And yet water (H_2O) puts fires out! All very puzzling. Let me try and explain.

Normally the electrons belonging to an atom stay close to their nucleus. So, what happens when a second atom comes along with its own nucleus and swarm of electrons? In some cases, the electrons of the first atom are so tightly bound to their nucleus they completely ignore the second atom. That's what happens with a gas like argon. It's very stand-offish. With other atoms it is not like that. Although their inner electrons are strongly bound to the nucleus, they have some loose ones on the outside. When the second atom comes along, the loose ones can get partly attached to the second nucleus. This is especially so if the electrons of the second atom are arranged in such a way that an extra electron can get in quite close to their nucleus. In fact, the electrons from both atoms can rearrange themselves in such a way that the two atoms get cemented together to form a molecule.

What happens if now a third atom comes along? It depends. Does the molecule that has already been formed still have any loose electrons on the outside? Are its rearranged electrons such that a loose electron belonging to atom number three can get in close to its nuclei? If so, the three atoms might form a bigger molecule; if not, the molecule becomes stand-offish – even though the individual atoms from which it is made are anything but stand-offish. The second case is a bit like a boy and a girl who are normally very outgoing and friendly. But one day the two of them meet, and fall in love with each other. From then on they always go around together and are no longer interested in anyone else.

That's what is happening with carbon dioxide. The two oxygen atoms are so attached to the carbon atom they don't want to have anything to do with anything else. So they no longer get involved in all the activities that single oxygen atoms usually do. As far as helping out with all the processes going on in our human bodies, they are now useless. The same goes for the hydrogen and oxygen atoms in water. Normally these atoms are so eager to link up with other atoms they can be dangerous and cause fires. But in water molecules they are already linked up; they are hanging on to each other so tightly, their behaviour changes completely.

 Why is the snow white? And why is it so cold? Why is snow so soft like cotton wool? And why did we start making snowmen?
Aimee

As I was just telling Miranda, water molecules jiggle about. They pull on each other, but also, because they can't sit still, they don't stay in the same place but slide over and under and around each other. That's why it is easy to stir a liquid. But

11

as the liquid gets colder, the jiggling gets less. Suddenly, when the liquid freezes, each molecule stays where it is, clinging to the same neighbours.

It's a bit like musical chairs. While the music plays, everyone keeps walking. But once the music stops, everything changes: you grab the nearest chair, sit in it firmly and hang on for dear life.

That's the kind of thing that happens when the temperature drops below the liquid's 'freezing point'. Suddenly the molecules lock on to each other and hold on grimly. That's when water becomes ice, or raindrops become snow.

With our own game of musical chairs, the rule is one person per chair. If you grab a chair, and someone arrives a fraction of a second later – too bad for them. But with the molecules' version of the game, sitting on laps is fine. That's part of the game. If we were to play their version, then someone would sit on your lap, then someone else would sit on *their* lap . . . and so on. You end up with a long line of people all sitting on someone's lap. And the same thing is happening on either side of you; you get several lines of people sitting on each other's laps. And each line is likely to be different depending on whether the individuals happened to sit down facing left, or right, or straight on.

That's how we get snow. The molecules latch on to each other in a regular way to form crystals. All the beautiful crystal shapes are slightly different from each other because of the particular way the

12

molecules latched on to each other.

So that answers one of your questions: snow is cold, because if it weren't cold it wouldn't be snow – the molecules would be jiggling so much you'd have a liquid; you'd have water.

Why is snow white? As we've seen, snow is made up of tiny ice crystals. These are solid and have lots of tiny surfaces. They reflect sunlight, in the same way as sunlight reflects off a window-pane. The whiteness can be quite dazzling on a bright sunny morning, just as light reflected from a window can dazzle you.

Why is it soft? The crystals of snow are made up of ice, and solid ice is hard. But the snow crystals are all spread out, with long thin spikes; there's plenty of space beween them. So, it is easy to scrunch them up together, packing the ice crystals together more tightly – which is what you do when you gather up some snow and squash it down to make a smaller, harder snowball.

As for who it was who made the first snowman, I guess it was the first child to wake up and find it had been snowing in the night.

 What are clouds made of? Why do they stick together?
Polly Glass

Let's start with puddles left after a rain shower. As the Sun comes out the water warms up. Its mol-ecules jiggle about. Some move so fast, they escape the pull of the others. They go drifting off to be among the molecules that make up the air. Normally we don't see them because a molecule is very, very tiny. If they bump into another molecule, they bounce off each other; they stay as a single molecule.

Now, the air near the ground is warm compared to that higher up. (The Sun's rays heat up the ground, and the ground warms the air next to it.) As you might have heard, hot air rises. This is because the molecules in hot air rush about like mad knocking other molecules out of the way, so they tend to be spread out. If the air is cold, on the other hand, then the molecules move about more gently; they can snuggle up closer to each other without getting knocked flying. This means a box full of air would have less molecules in it if the air was hot; more if the air was cold. And that means the box of hot air would be lighter than the box of cold air. That's how hot air balloons work. The hot air inside the balloon is less dense than the cold air around it, so it rises.

So, as I was saying, the water molecules that evaporate from the puddle join the air close to the ground, and this rises. As it does so, it cools. Now the molecules jiggle about less. They don't bump into each other as hard, and they can now stick together. We have the beginnings of a water droplet. Eventually the droplet grows big enough to be seen. This goes on wherever the air is cold enough, so we have lots of water droplets. And that is what a cloud is: it's a collection of water droplets.

Actually getting a droplet to *start* growing is not quite as simple as I have said. Clouds form better if the air is dusty. At the very beginning, the water molecules stick to the dust particles, and only after the droplet has grown a little, do they start sticking to each other.

As for what holds the clouds together, that isn't a problem. The shape of the cloud is just the shape of the region of the air cold enough to form droplets. So the cloud isn't really held together at all. There's no need for glue – just as you don't have to glue sunbathers together. You find sunbathers together on the beach because that is where it happens to be warm!

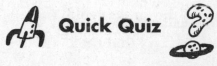 **Quick Quiz**

A friend of mine makes clouds. Would you reckon she must be a scientific genius?

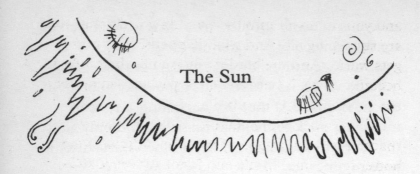

The Sun

How was the Sun made? Is it all gas and fire?

Hal Jarret

Yes, you are quite right: the Sun is a great ball of fiery gas. Like everything else, it is made of atoms. Except you don't need ninety-two types to make a Sun; just two kinds will do. These are the lightest atoms; they are called *hydrogen* and *helium*.

What happened was that there was this huge cloud of hydrogen and helium gas floating in space. Everything in the Universe pulls on everything else with a force called *gravity*. (At this moment you are being pulled down into your seat by the gravity force of the Earth.) The same was true of the atoms that made up this cloud; each atom in the cloud pulled on every other atom. Atoms are not very heavy so their gravity force is very gentle. But over a long period of time this force slowly squashed the cloud smaller and smaller.

Ever had a flat tyre? You get the bicycle pump

16

and you pump air into the tyre. As you do that you are squashing air. And what happens? The pump gets warm. And the harder you pump, the hotter it becomes. This is because when you squash air, or any other gas, it heats up.

That is what happened to the cloud of gas in space. The smaller it got (because of gravity) the hotter it became. In the end it got so hot, it suddenly caught fire.

16 How long has the Sun been in space? (Hope you get it right.)
Jamie Oilk (aged 9)

The Sun caught fire (it was 'born') about 4,600 million years ago – and the fire has been burning ever since.

How can we grasp what 4,600 million years is like? One way is to imagine collecting calendars (the sort that have a new page for each month). You start saving them way back when the Sun was born, and you buy one every year. If you did that, by now, the pile of old calendars would be 10,000 kilometres high! That's almost the distance from one side of the Earth to the other.

17 Where does the Sun get its heat from?
Paul

In the early days, the Sun warmed up by having its gas squashed down. But since it 'caught fire', it has got its heat another way – from the atomic nuclei.

What happens is that the Sun gets so hot (millions of degrees), everything jiggles about like crazy. That's what we mean when we say something is 'hot'. We mean that the bits it is made from vibrate and move about with lots of energy. In the Sun everything is banging about so much, the electrons get knocked out of their atoms. The electrons and nuclei go charging about like mad, here and there in all directions at once, bumping into each other.

When nuclei of hydrogen and helium crash into each other like this, they sometimes stick together. When this happens, they give out energy in the form of heat and light.

Why do they give out energy? It's a bit like two people who start off living in separate houses. They each need to use gas and electricity to keep themselves warm, heat the bath water and do their cooking. But if they decide to live together, they cut their fuel bills. They can sit in the same heated room instead of separate ones, watch the same TV set, cook both their meals in the same heated oven, and so on. That means there is now more gas and electricity (energy) that can be used in other ways.

The same is true of the nuclei: when they get together, they don't need as much energy as when they were separate. So when they bang together and stick, out comes the energy they no longer need. And that's the energy that keeps the Sun burning.

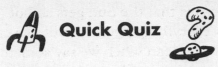

Quick Quiz

Nuclei are so tiny you wouldn't expect much energy to come out of just two of them sticking together. So why is there so much heat in the Sun?

What would happen if the Sun got too hot? I think it would explode and then all the other planets including the Earth would get too hot and explode. But I would really like to know the proper answer.
Rebecca Brown (aged 10)

The Sun is like a giant nuclear bomb. It gets its energy in exactly the same way as a nuclear bomb does – through nuclei sticking (or fusing) together. But the great thing about the Sun is that it is a bomb going off s . . . l . . . . o . . . . . w . . . . . . l . . . . y.

It is really quite amazing how the Sun manages to do this – how it is able to feed the hydrogen fuel into its fire at just the right rate to keep the fire burning at a steady rate.

Mind you, this cannot go on for ever. It's not that the Sun is in any danger of suddenly blowing up. But sometime in the future, it will slowly start to swell up. When it does, the Sun will turn red and almost fill the whole of our sky. The surface of the

Earth will become mega hot and all life will be burned to a frazzle.

That's the bad news. The good news is that you and I, and our children, and grandchildren, and our great grandchildren, and great great etc . . . . will all be dead and gone long before this happens! The Sun is due to stay more or less the same as it is for about another 5,000 million years. So you can't use 'the end of the world is coming' as an excuse for not doing next week's homework. Tough!

How do you know what the Sun is made of if you can't touch it?
John

You're right, we can't touch the Sun. So, we have to make use of the light we get from it.

When you heat up atoms in a laboratory, they give out coloured light. For example, heat up atoms of sodium and they shine with a bright yellow light. (This is the light you get from yellow sodium street lights.) If you heat up another type of atom, neon, you get the pinkish light of neon strip lights; and so on. Not only that, if the sodium atom is cold rather than hot, it will swallow up any yellow light that is shone on to it, and neon will swallow up pink. So each type of atom has its own special likes and dislikes when it comes to colours.

This special mix of colours becomes a kind of atomic fingerprint. In fact, you can turn all this into a detective game. If I heat up some unknown mixture of atoms, can you tell me what atoms I've got simply from looking at the colours given out? That is the kind of detective problem set by the Sun. Instead of just saying the Sun is a sort of yellowy white, and leaving it at that, you measure very, very carefully exactly what colours are being given out and getting swallowed up. And that's how scientists learnt that the Sun has been formed mostly out of the two lightest types of gas: hydrogen and helium – and they don't have to burn their fingers finding out!

How does the Sun hurt our eyes if it is so far away?
John Baldry (aged 11)

The Sun is a long way from us. If a spacecraft were to fly to the Sun from the Earth, and it went at the same speed as a jumbo jet, it would take twenty years to get there. It is the distance that makes the Sun look small.

But in fact it is big – so big, a million Earths could fit inside it. Because of the distance, the Sun's disc covers only a tiny fraction of the sky. We get heat from only that small part of our sky; that's why the Earth is nice and warm instead of baking hot.

BUT the *brightness* of the Sun's small disc is more or less the same as if the Sun were right up close to us. This is because it travels most of the way through empty space, and there is nothing to block its path to us. So, although the disc is small, it is ever so bright – dangerously bright. After all, you

21

can burn holes in wood with the Sun's rays using a magnifying glass. (So, what do you think you would be doing to the back of your eye if you looked directly at the Sun through the lens that makes up the front part of your eye!)

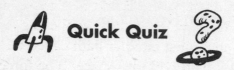

Quick Quiz

Why is it safe to look at a white surface, like a page from this book, even when the Sun's rays are shining directly on it?

If the Sun goes away for ever there will be no warmth or light or heat. If there was no light how would the plants, trees, and flowers grow? I hope the Sun does not go away.
David

As you say, without the warmth of the Sun, there would be no life on the Earth. In olden days, people got very worried that the Sun might go away and not come back. After all, every evening at sunset it had the habit of disappearing, and no one knew where it went to at night. Who could be sure it would be back next day? Even more alarming were eclipses. These happen when the Moon comes between us and the Sun, and so blocks out the

Sun's rays for a short while. Fancy the Sun disappearing in the middle of the day without warning!

Nowadays we don't have to worry about such things. We know why the Sun sets and rises. We also understand why eclipses happen and we can predict exactly when to expect them.

But *could* the Sun and Earth ever get separated? Not really. We now know that the Earth goes round the Sun in an almost circular path called its *orbit*. What stops it flying off into space? The force of gravity between them. If you try escaping from the Earth by jumping, the Earth's gravity pulls you back again. It is the same with the Earth and the Sun. Because the Sun is so massive, it has a huge gravity force, and it is this force that makes sure that the Earth will always stay close to it. And a good thing too!

16 Why does the Sun not move?
Abdul (aged 11)

One of the interesting things about being a scientist is when you find out that you've been wrong! At first it was thought that the Earth was flat. But, in fact, it was round. Then it was thought the Earth stood still and the Sun went around it. But that again was wrong. It was the Sun that stood still and the Earth went around it. The Earth is speeding round the Sun at thirty kilometres per second; that's over a hundred times faster than a jumbo jet.

(Doesn't feel like it though, does it?)

Now we're wrong again! The Sun *doesn't* stay still. It belongs to a *galaxy*. Galaxies are great collections of stars. Like the other galaxies, our Galaxy (spelt with a capital 'G') is rotating; it is spinning slowly about its centre like a giant roundabout. So the Sun is slowly orbiting around the centre of the Galaxy. When I say 'slowly' I mean it takes a long time to get right round – 200 million years to make one complete turn. That's because it has a long way to go. The Sun (and its planets) are actually moving quite fast: 230 kilometres per second around the centre of the Galaxy. That's ten times faster than the Earth's speed in orbit around the Sun. Phew!

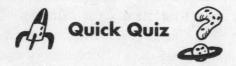 **Quick Quiz**

To find out how fast the Earth is really moving, we need to know how fast it is going relative to the Sun, and how fast the Sun is going relative to the centre of our Galaxy. But we are not home and dry yet. We need to know how fast one other thing is going. What do you think that might be?

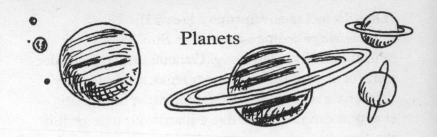

Planets

 How many planets are there in space?
Sarah

Nine of them. Starting closest to the Sun we have
Mercury, Venus, Earth, Mars, Jupiter, Saturn,
Uranus, Neptune and Pluto.

They all go round the Sun, each one in a bigger
and bigger path, or orbit. Earth goes round in one
year – that's what we mean by a year. Pluto's year is
the longest: 248 Earth-years. Mercury goes round
in eighty-eight Earth-days – a quarter of the time.
So, if you are ten years old, say, and someone was
born on Mercury the same day as you, they would
reckon themselves to be forty by now – forty
Mercury-years old, that is. (Not that anyone ever
does get born on Mercury – being so close to the
Sun, it is far too hot.)

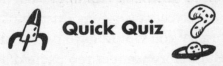 **Quick Quiz**

If you had been born on Pluto instead of Earth, how
long would you have to wait between birthdays?

18 How far is Mercury from the Sun?
Andrew Buchanan

Although it is the closest of the planets to the Sun, it is still 58 million kilometres from the Sun.

How far is that? Imagine a length of string wrapped round the Earth's equator. Now stretch it out straight. It would take 14,500 such pieces of string, placed end to end, to reach from Mercury to the Sun (37,500 to reach from Earth to the Sun).

From the surface of Mercury, the Sun would look three times larger than it does from the Earth; that's why during the Mercurian day it is boiling hot there and no life would be able to survive. In contrast, at night time when you are facing away from the Sun, it becomes unbelievably cold. This is because Mercury does not have any atmosphere or clouds to act like a thick blanket keeping the warmth in.

Another odd thing about Mercury is the length of its day. A 'day' is how long the planet takes to spin on its axis from noon one day until noon the next. The Earth spins 365 times in the time it takes to complete an orbit, i.e. it has 365 days per year. Because Mercury spins so slowly, and its orbit is that much shorter, it completes two orbits round the Sun for each time it spins on its axis; in other words it has two years every day. Just think: if you lived on Mercury you would have two birthdays every day! But before you get too excited at the thought of all those presents, remember what I said

about it being hot. Do you really want to be fried
like a piece of burnt crispy bacon – and then for
good measure, end up at night as a frozen piece of
burnt crispy bacon?

 **Why are the planets all different sizes?
And why is there life on the world and
not on Jupiter?**
Gregory Roostan **(aged 8)**

They are very different in size. Jupiter is the biggest.
Its diameter (the distance from a point on its surface to the point on the exact opposite side) is
eleven times bigger than that of the Earth. Pluto, on
the other hand, is the titchiest – less than one-fifth
of the Earth's diameter.

Not that even Pluto is all that small. If you visited
Pluto and decided to walk right round it, it would
still take you a year to get back to your starting point
– even if you walked twelve hours every day. As for
doing a walking tour of Jupiter, that would be a real
marathon; it would take you your whole lifetime.

Actually, be warned. It is not a smart idea to try
to walk on Jupiter at all. It has no surface to walk
on! It is really nothing more than a great ball of gas.
If you parachuted into Jupiter, you would sink
through the atmosphere. This would get denser and
thicker the deeper you sank. After a while, it would
become so thick it would be more like a liquid than
a gas. The liquid would then become as dense as
treacle. At that point you would stop falling and

you'd just float. But you wouldn't be standing on a surface of any kind. So it is not difficult to see why Jupiter is unlikely to support life. The same is true of all the really big planets: Jupiter, Saturn, Uranus and Neptune.

But, although some planets are mostly gas, and the others are rocky, they were all formed in the same way. At the time gases and dust clouds collected together to form the Sun, little eddies or whirlpools were set up outside the centre (like you sometimes get when the water is going down the bath plughole). Instead of getting sucked into the newly formed Sun itself, they stayed outside going in orbit round the Sun. They each squashed down (because of gravity) to form the planets.

For the inner planets close to the Sun, the light gases were blown away by the hot wind coming from the Sun, leaving the collected dust which now formed the rocky planet. Further out, the planets like Jupiter were able to hold on to their light gases – that's why they are still mostly gas.

So, why are the planets different sizes? Instead of me just telling you the answer . . .

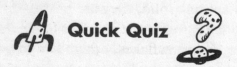

Quick Quiz

. . . why not have a go at answering it yourself, now that you know what I have just told you?

28

While I was watching a space programme with my sister, they showed a map of space. I stared at Saturn and thought why does Saturn have rings?
Joseph Stewart

As a boy, I lived for a year with my Uncle Bill. (It was during World War Two, when it was too dangerous for children to live at home in London because of the bombing.) He had a wonderful telescope. It was his hobby. He loved looking at the stars, and showed me all sorts of wonderful things through it. It was Uncle Bill who first got me interested in 'what was up there'.

But the trouble with star-gazing is that the best time for doing it is when it's cold and clear. Some years later, Uncle Bill became very old and weak and his wife wouldn't allow him out on cold nights anymore. To my delight, he gave me his telescope. It is still my pride and joy.

Without doubt, the greatest thrill I ever had with my telescope was the sight of Saturn and its beautiful rings. I know exactly how you feel about them, Joseph.

So, what are they? They are very flat and thin. But they are not solid; they are not rigid. They are in fact made up of a vast number of pieces of ice. Some are like snowflakes, others like dirty snowballs, and some are as big as a snowman's body. They slowly move round and round the planet, much like the Moon goes round the Earth. In fact,

you can think of these icy bits as very tiny moons.

We now know that Saturn is not alone in having rings. Jupiter, Neptune and Uranus also have some, but they are very thin and hard to see even when they are photographed from close up by a space probe.

21 What is a comet?
Quentin Chan (aged 9)

Comets are dirty snowballs, five to fifty kilometres across. Like the planets they orbit the Sun. As astronomical objects go, they are not very important. So why do they cause so much fuss when it is announced that one is on the way?

First of all, let us think of their orbits. All objects that orbit the Sun do so in elliptical paths. An ellipse is like a circle that has been squashed in one direction. The Earth follows a roughly circular orbit (hardly any squashing), so it does not change its distance from the Sun much. A comet, on the other hand, tends to follow a more elongated path. That means it comes much closer to the Sun at certain times of its 'year' than at others. When this happens, it warms up. Some of the ice melts and gases, vapour and dust clouds are given off. These float away from the heart of the comet to fill up a huge amount of space. These dusty gases and vapours are lit up by the Sun and are seen from the Earth as a fuzzy cloud. The light and the wind from the Sun are so intense at these close-in distances that the cloud is blown away from the head of the comet in a direction directly away from the Sun. So the cloud stretches out into space like a long 'tail'. And that is why they can look so spectacular.

After its closest approach to the Sun, the comet continues its journey. This now takes the heart of the comet further and further away from the Sun. It becomes cold once more, and so no longer gives off its vapours. It disappears from view as its path takes it back out into the further reaches of space. Meanwhile astronomers calculate from the shape and size of its orbit how long it will be before it returns once more.

The Earth

22 **Everything has a beginning. How was the Earth made, and when?**
Kathryn Brown

I have already pointed out that the Earth is a planet like the others. So, it was formed out of a whirlpool of gas and dust outside the Sun. We think it formed at the same time as the Sun and the other planets, i.e. 4,600 million years ago.

How do we know that? Because of radioactivity (pronounced 'radio-activity', except that it has nothing to do with the activity of listening to Walkmans). What happens is that some very heavy atomic nuclei are too big and wobbly for their own good. After a while, bits get thrown off, or fall off them – leaving a somewhat smaller, more sensible-sized nucleus. When a nucleus slims down in this way, we say it has 'decayed'; it has radioactively decayed. For any particular type of nucleus, half of them decay in a certain time called its 'halflife'. For example, the halflife might be 1 million years. That means, if you start off with sixteen oversized nuclei, after 1 million years you will be left with half of them, namely eight. So that gives you eight nuclei that are still oversized, plus eight new sensible-sized nuclei. After *another* 1 million years, half of the remaining eight oversized nuclei will also have decayed. That means that after a total of 2 million years, we shall have only four nuclei that are still oversized, and four more sensible-sized ones to add to the eight we already have, giving a total of twelve sensible-sized ones.

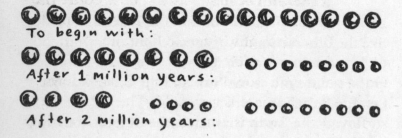

Can you see how this gives us a way of working out how long a collection of nuclei has been around?

By examining the dirt of the Earth, and counting how many oversized nuclei we have compared to sensible-sized nuclei, we can work out how old the collection is. That is how scientists are able to work out that the Earth is 4,600 million years old. And because they believe the Sun was formed at the same time as the Earth, that must be how old the Sun is.

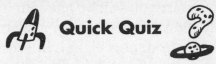

Quick Quiz

Suppose I have a collection of nuclei like those we have just been talking about. In this collection, there are two oversized ones and fourteen sensible-sized ones. Can you tell me how long my collection of nuclei has been around?

If the Sun is so far away, how come the middle of the world is hot? You would have thought it would be cold.
Melanie (aged 10)

Good point. You would expect it to be cold down there. After all, space is very cold. The reason the surface of the Earth is warm is that it gets heated up by the Sun during the day; at night it rapidly loses heat to space. I don't have to tell you how quickly things cool down in winter on a clear night once the Sun has set.

But in fact, if you go down a mine, the deeper you go, the *warmer* it becomes. As for the centre of the Earth, it is very, very hot – so hot, it melts rocks! And that is without it *ever* seeing the Sun down there. So, what's going on?

It's all to do with those atomic nuclei again. Remember how the Sun got its heat? In question 11, I told Paul how light nuclei banged into each other and fused together to form somewhat heavier

35

nuclei. As they did this, some heat energy was given out. I said the Sun was a kind of nuclear bomb going off slowly.

Well, there is another kind of nuclear bomb. It's going off quietly *under our very feet!* Remember the big oversized nuclei throwing off bits to become more sensible-sized? Those changes to the nuclei also send out heat energy. Whereas the Sun's energy comes from nuclear *fusion* (the fusion, or sticking together, of light nuclei), the energy at the centre of the Earth comes from nuclear *fission* (the fission, or splitting up, of heavy nuclei).

But 'Ah!' you're thinking to yourself. 'He told Paul that you save energy when you live together. That's how he explained energy from fusion. Now he is saying the *opposite* – that you save energy when you split up. He can't have it both ways.'

Oh no? Who says? If you have a house with a teenager living in it – one who is always leaving the lights and the TV on, it is very easy to save energy when they leave home! Some people save energy when they get together, others when they split up. The same goes for nuclei. Some nuclei give out energy through fusion; others by fission.

Of course, you don't get much energy out of just one nucleus splitting up. But as we saw with fusion, when you get lots and lots of nuclei all doing the same thing at the same time, it adds up.

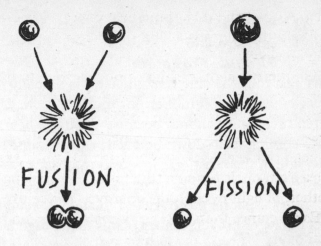

FUSION

FISSION

What happens to all this energy being produced deep down in the Earth? It makes its way to the Earth's surface. From there it is able to escape into space. But this takes a long time; the heat has to travel thousands of kilometres to get to the surface. Because of this, the temperature builds up – to the point where it melts the rocks.

It's quite a thought that the solid-looking land we walk on is not actually solid all the way down. It is in fact just a thin floating crust – a bit like the cool skin that forms on the surface of hot custard!

(Which reminds me: I *love* the skin on custard; do you? I always had the skin. But then I got married. To my horror I discovered that my wife's mother also loved the skin on custard. Can you imagine: two grown-ups fighting over who was to get the skin?!)

What happens in rock before an earthquake?

Thomas Compton

I have just told Melanie how the land is actually a thin crust floating on molten (or melted) rock. This molten rock churns about because of all the heat that is coming from down below. It rubs against the underside of the crust on top – and that makes the crust slide about. Cracks form in the crust, so it ends up as separate pieces called 'plates'. It's a bit like a jigsaw of broken dinner plates fitted together.

One of these cracks runs down the west coast of the USA. The city of San Francisco is built mainly on one of these plates, and that of Los Angeles, about 600 kilometres south of San Francisco, on the other side of the crack.

Interesting things happen near these cracks, or 'faults' as they are called. The churning stuff underneath one of the plates might be slowly moving that plate in some direction, while that under the next-door plate is moving that one in a different direction. This is happening in California. Los Angeles is advancing towards San Francisco. Not that the rail fare between the two cities shows any signs of being reduced year by year. The gap is closing at the rate of only five centimetres per year.

But what, you ask, has this to do with earthquakes? It's all right; I haven't forgotten the question. It's just that sometimes the answer takes a while coming. Here it is:

If we had a giant oilcan, we could keep the gap between two plates nicely oiled all the time – so the two could slide smoothly past each other. That way all would be well; there would be no earthquakes. But it is not like that. For much of the time the plates get caught up and stuck together at the boundary between them. While the main parts of the two plates continue to move past each other at a steady rate of a few centimetres per year, the edges aren't budging at all; they are being left behind. After ten or twenty years of this, they might be lagging by a metre. Strains are set up in the rocks on either side of the divide. Another ten or twenty years, and they are now lagging by a couple of metres. The strains are increasing. Another ten years . . . It can't go on like this. It is like trying to stretch a rubber band. You can go just so far, then . . .

TWANG!

The twang is the earthquake. A point is reached where the strains are so great that the rocks split, the plates free themselves from each other, and suddenly jerk forwards in the direction they have been trying to go in. There is a shudder that ripples outwards from the fault causing damage to buildings. Often people get killed.

What happens then? The plates get caught up again, and the whole thing is set to repeat itself once more. There is then another earthquake – near the same place – in a few years' time. Then another, and another . . .

If it is known that some places, like California, are dangerous (San Francisco was completely destroyed by an earthquake in 1906), why do people live there, you might ask? Why not be sensible and go and live in the middle of a plate where there is little chance of an earthquake? Good idea. I used to think people who lived in California must be thick or something. But then, when I was a young scientist, I got the offer of a very interesting job. It was at the world's leading place for doing my type of nuclear physics. The trouble was it was right next to San Francisco! What should I do?! (What would *you* have done?) I thought about it – and decided to go. I stayed there a whole year. Does that mean I'm thick? Probably. Thick – but lucky; it was a wonderful year.

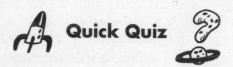 **Quick Quiz**

You want to go from San Francisco to Los Angeles, but you haven't got the rail fare. You decide to wait until the movements of the Earth's surface bring the two cities opposite each other. That way you will be able to just walk across. How long will you have to wait?

Why do volcanoes explode? I would like to know this because I like the volcano colours when they explode.
Sarah Bartholomew

Volcanoes occur where there is a weak spot in the Earth's crust. Over a period of time, pressure builds up in the molten rock down below. This goes on until the crust can no longer keep the lid on. Suddenly, the weak point gives way and the molten rock comes shooting out. That is how the explosion happens. As the molten stuff comes out, it cools down to form ash and solid rock, which then gets dumped all around the hole. Each time there is an eruption, the mound of ash and rock grows. In the end, there is so much of it, it becomes a mountain – a mountain with a hole at the top, where the stuff is still coming out.

These weak points often occur at the boundary between two plates. So what with earthquakes, and now volcanoes, it really is a good idea not to live in such places!

As you say, volcano explosions can be very beautiful – like a gigantic free firework display. But they are very dangerous and destructive. In 1883, a particularly large one occurred at Krakatoa, on an island in the Indian Ocean. It killed 40,000 people. The sound of the explosion could be heard 5,000 kilometres away.

At night when the Sun goes down, I wonder where it goes. Does it go into the sea or does it go to Australia? If it is not any of those, tell me the answer anyway.

Elliot Wright (aged 9)

For a long time everyone thought that the Earth was flat. Each morning the Sun would rise up, pass overhead, and then disappear into the distant ground, or sea, in the evening. As we saw in Question 15, when I was writing to David, this was very worrying. What if the Sun decided not to come back again the next day?

But then it was discovered that the Earth was *not* flat. It was round, like a ball. As for the Sun, it journeyed in an orbit around the Earth. At night time it didn't dive into the sea; it simply went round the corner and shone on a different part of the Earth's surface, before reappearing the other side.

This certainly seemed to explain things nicely. Except that this idea was also wrong!

You know how when you sit on a roundabout in the park, it looks as though everything – the trees, the park benches, the prams – are all whizzing round you. But they're not really. It is not *they* who are moving; it is *you*; your roundabout is spinning. Well, that's how it is with the Earth. The Sun is *not* moving round us; it is the Earth that is spinning. It spins about a line (or 'axis') passing through the Earth from the North Pole to the South Pole.

All the planets spin, but at different rates. So their 'days' are different from ours. Jupiter's day is just under ten hours; Mercury's is 176 Earth-days.

So, the answer to your question is that at night time the Sun goes to Australia – or better still, it is the turn of Australia to spin round to face the Sun.

27 Why don't we notice the world spinning?

Naomi Durston

You might wonder why it took so long for people to twig that the Earth was spinning like a roundabout. After all, if you ride a roundabout all day, you know it's you that is moving – you end up dizzy! So why don't people at the North and South poles end up

dizzy? And why don't those living along the equator get flung off into space? You've guessed it. It's because the Earth spins so slowly. No one gets dizzy on a roundabout that takes a whole day to do one turn.

 If it takes a year for the world to go round the Sun, why do we have leap years?
Sophie Fowler

We have seen how a year is how long it takes the Earth to go right round the Sun in its orbit and get back to its starting point. A day is how long it takes for the Earth to spin on its axis between noon one day and noon the next. But there is nothing to say that the Earth has to spin on its axis an exact number of times while completing one orbit round the Sun.

In fact, the Earth's year is 365¼ days long. This means that, having started its orbit at a particular point in space on 1st January, after 365 days (what we normally call a 'year' – from 1 January to 31 December) the Earth is not quite back to where it was at the start. After two 'years' it is lagging by twice as much; after three 'years', three times as much. After four 'years' it would take the Earth four quarter days to make up the difference. But four quarter days is, of course, one whole day.

Which is where the neat trick comes in. Every fourth year we announce that there is an extra day to the 'year' – 366 instead of 365. During that extra day the Earth is able to catch up. By the next 1 January, it is ready to start the new 'year' from the proper starting point of its orbit. The extra day is tacked on to the end of February – to give 29 February. (Don't ask me why; I reckon it should have been 32 December.) Anyway, it's a good idea (provided you don't have your birthday on 29 February). Without these *leap years*, as they are called, the calendar would gradually slip. Winter would eventually come round in July and summer at Christmas time. (Which is something the Australians have always had to put up with; but that's *their* problem!)

PS Actually that's not the end of it. The Earth does not orbit the Sun in *exactly* 365¼ days – it's a little bit less than that. So, even after a leap year, the Earth isn't *quite* back to where it ought to be; it is now a teeny bit *ahead*. So to make up for this, every hundredth year (those ending in 00 like 1800 and 1900) instead of having a 366-day leap year, they have an ordinary 365-day year.

PPS Even that isn't the end of it. That skipping of a leap year every hundredth year, still leaves the Earth just a teeny weeny bit out from where it ought to be. So to make up for this, every *four hundredth* year (like the year 2000), instead of having an ordinary year, they have a leap year.

PPPS By now, I bet you're wishing you hadn't asked!

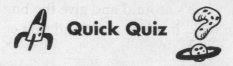

Quick Quiz

I said that every four hundredth year the Earth was still a bit out from where it ought to be. What did I mean by that: was it a bit ahead, or a bit behind? (No need to guess; look at the correction they make, and you can work out the right answer.)

 How do we know what's inside the Earth?
Laura Sedgwick

The obvious way is to dig holes and see what you come up with. But this only scratches the Earth's surface. The deepest mine is about three kilometres and the deepest borehole about fifteen kilometres – which isn't much compared to the 6,370 kilometres it takes to get to the centre of the Earth.

A better way is to let the Earth bring up its own inside. This is what happens with a volcano. Lots of hot rock comes pouring out during an eruption. This tells us that the inside of the Earth is very hot. We can also examine the kind of rock brought up (once it has had a chance to cool!).

But there is another way. Suppose the postman delivers a present – a brown paper parcel. You are dying to open it and see what's inside, but you're not allowed; you have to wait until Christmas Day, or your birthday, or whatever. What do you do? You wait until no one's around and give the box a good shake. With luck, it might rattle and that could be a clue. For example, if it does rattle, it can't be boring socks and vests.

Scientists can do the same to find out what's inside the Earth. They give it a shake. Actually, they don't have to. From time to time, the Earth gives itself a shake: an earthquake. The violent movements set up ripples, and these spread out through the Earth's interior. We call them 'earthquake waves'. They can be picked up at different points around the Earth's surface. The paths they take depend on what they are travelling through. They curve about and sometimes get reflected when one type of rock meets up with another.

When there is an earthquake somewhere, scientists all round the world study the waves when they reach their recording equipment. They then compare their results – what kinds of ripple they found, how strong they were, and how long they took to arrive at their particular point on the surface. Then comes the detective work. From all their results they work out what the inside of the Earth must be like. The picture they have come up with is this:

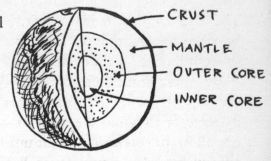

CRUST
MANTLE
OUTER CORE
INNER CORE

The inside of the Earth is a bit like an onion with several different layers. On the outside is the skin or solid *crust*. This is thick where there are continents (it can go ninety kilometres down under mountains), and can be as thin as five kilometres where the oceans are.

Under the crust is the *mantle*, which is made of the kind of stuff coming out of volcanoes. Then 2,900 kilometres down (about half-way to the centre) comes a big change: we reach the *outer core*. This is liquid. We know this from what happens to a type of earthquake wave (called an S wave). S waves shake from side to side as they go along (in the same way as you can send a wave along a length of rope by shaking it). This type of wave cannot pass through liquid.

This sort of wave never reaches the opposite side of the Earth to where the earthquake happened. This must be because it can't get through the core in the middle; the core casts a kind of 'shadow'. And the size of the shadow tells us how big the liquid core must be.

Then, at a depth of 5,150 kilometres, we reach the surface of a solid *inner core*. We know of this because of what happens to the other main type of earthquake wave, the P waves. P waves are pressure waves (the vibrations are along the direction of motion rather than side to side). They can travel through both liquid and solid.

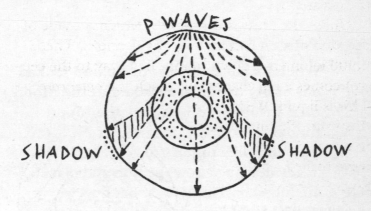

The inner core is slightly smaller than the size of the Moon. We think it is made of iron because it has to be very heavy. We know how heavy the whole Earth is, and this tells us that the stuff at the centre must be about 4 times heavier than the lighter stuff we find in the crust. This points to it probably being mostly iron.

Finally, after going down for 6,370 kilometres, we reach the centre of the inner core – the centre of the Earth.

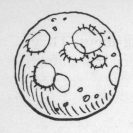

The Moon

When I went to bed I got up and looked out of my window. I looked at the Moon then I wondered how was the Moon made?

Steven Jones (aged 10)

I have already described how the Sun and planets were made (question 19). Remember how we started with a big swirling cloud of dust and gas. Most of it got drawn into the centre to form the Sun. But smaller eddies, and whirlpools, formed outside the centre. These settled down to become the planets going round the Sun.

One idea about how the Moon was born is that a very little whirlpool formed in the gas and dust outside the Earth. This squashed down and ended up as the Moon going round the Earth. That is certainly how we think the other planets got at least some of their moons (Mars has two, Jupiter sixteen, Saturn eighteen, Uranus fifteen, Neptune eight, and Pluto one).

But that is not the only way to get a moon. In the early days, soon after the Sun and the nine planets formed, there were lots of rocks flying about. They were, if you like, very, very tiny planets. It is thought that some of them came too close to one of the big planets, and got captured in orbit about it. From then onwards they became one of the planet's moons.

But today most scientists think that *our* Moon wasn't formed in either of these ways. They think that, soon after the Earth formed, it was hit by one of these rocks flying through space. A huge chunk was knocked out of the Earth. This came out as a great 'splash' of molten rock. It went into orbit around the Earth, cooling and settling down to give us the Moon.

39 Why does the Moon have so many craters on it?
George (aged 10½)

For a long time people used to argue over whether the craters on the Moon were due to extinct volcanoes or the impact of those rocks flying about space I was talking about just now. We now know it was mostly the rocks that were to blame. Planet Mercury looks much the same as the Moon; it too is pitted all over with craters.

Now you might be wondering why these two came in for such a battering and not the Earth – after all there aren't many craters on Earth. Well, it's not that we escaped bombardment. (Recall how the Moon was made through such an impact.) The point is that we have an atmosphere; we have winds and rain. In short we have *weather*; the Moon and Mercury don't have weather. And it's weather that wears down the sides of the craters and fills in the holes. It takes a long time, but so what? There has been a lot of time since the craters were formed. A few still survive today; some are very large and obvious, so we know that the Earth did indeed have to take a battering just like the other planets did. Fortunately things have calmed down a lot since the violent early times of the formation of the Solar System. By now most of the flying rocks have been gobbled up by the planets. But from time to time, there are still some that come flying through our atmosphere from outer space and crash on to the surface of the Earth. These rocks are called *meteorites*. Mostly they are small and do very little damage. But there are some big ones out there still, so keep your fingers crossed!

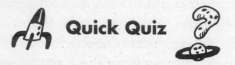 **Quick Quiz**

Do you happen to know what all this talk about meteorites hitting the Earth has got to do with dinosaurs?

32 Why does the Moon change shape overnight?

Kerry

The Moon isn't like the Sun; it doesn't give out any light of its own. So, how can we see it? By the light from the Sun reflecting from its surface. Sunlight is white, and the rocks on the Moon are a light greyish colour. The great flat plains caused by molten rock bubbling up from down below and spreading out are somewhat darker. That's why the Moon is blotchy – how it ends up looking like a face: 'The Man in the Moon'.

The phases of the Moon (whether it looks like a thin crescent, or a half moon, or a full moon) depends on the angle at which the Sun's light is striking the surface. If, for example, it is the right-hand half that is lit up, then the Sun must be over to your right; if it is a full moon, the Sun must be directly behind you. It takes about a month for it to go through its full range of 'shapes'. On a clear night with a crescent moon, if you look hard you can see the rest of the Moon making up the complete disc – the part that is in shadow from the Sun. How? By light coming from the Earth, and being reflected back to us by the darkened part of the Moon! Earthlight is to a Moonling what moonlight is to an Earthling. Moonlings standing on the Moon at night-time, see above their heads a shining Earth, lit by the Sun. So even though it's night-time on

53

that part of the Moon, the Moonlings can still find
their way about with the help of faint earthlight.

PS *Moonlings!?!* Since when have there been . . .?

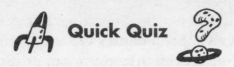

Quick Quiz

Do you know why it takes about a month for the
Moon to go through its various 'shapes'?

**I am interested to know why we get
eclipses of the Moon.**
William Palmer (aged 7)

We have seen that the Moon shines by the light it
receives from the Sun. But what if something got in
the way of the Sun's rays? It would be put in
shadow. And that is what sometimes happens; the
Earth blocks the path of the Sun's rays, and the
Earth's shadow falls on the Moon. Even so, you can
still see the Moon dimly. Why is that? Well, imagine
you are standing on the Moon. What do you see
during an eclipse? You would see the disc of the
Earth gradually blot out the Sun. When the Sun
was completely blotted out, you would see the
Earth as a dark disc, but with a thin illuminated
ring around it. That would be due to the light still
being scattered through the Earth's atmosphere. It
is that light that dimly falls on the Moon during an
eclipse and allows us here on Earth to see it.

As for an eclipse of the Sun, that is when the Moon gets in the way of the Sun's rays coming to Earth. A total eclipse is when the disc of the Moon completely covers the disc of the Sun.

 If the Moon comes out at daytime, why doesn't the Sun come out at night?
Matthew (aged 5)

The Sun shines all the time. It gives out so much light it turns everything into day. When we talk about 'daytime' we actually mean that time when the Sun is up. If ever the Sun decided to come up in the middle of the night (it can't – but just suppose it did), it would straightaway shine so brightly it would turn night into day!

Not so the Moon. The Moon gives out only a faint light. If it comes up in the middle of the night, so what? It sheds its gentle moonlight, but that is not enough to turn night into day. As for the day-time, it can come out, and again it doesn't make any difference – it's still just normal daytime.

So, although the Sun is big and important and makes everything daytime when it is around, I reckon the Moon has more fun. Like us, it can enjoy the evenings and the restful night-times as well as the busy daytimes.

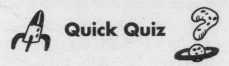 **Quick Quiz**

Can you get an eclipse of the Moon at those times
when the Moon is visible during the day?

Even Quicker Quiz
Why can't you see an eclipse of the Sun at night
time?

Stars

35 Why do stars glow?
Kristina

You know why the Sun shines. Remember the little atomic nuclei banging into each other and giving out energy (question 11). Well, it is the same with stars; they too are powered by nuclear energy. In fact, stars are suns. Each star is about as big and powerful as our Sun! I know they don't look it. That's because they are so far away.

To get some idea as to the distances of the planets and stars, imagine a cat called Mercury curled up on a rug thirty centimetres from a gas fire (which we'll pretend is the Sun), and an Earthling is stretched out on the sofa one metre from the fire. Then on that scale, the poor old dog, Pluto, would be forty metres down the road. As for the nearest star, that would be 250 kilometres away! No wonder they look tiny.

I was lying in bed and looked out of the window and I thought: Why do the stars come out at night?
Sarah Jarvis (aged 9½)

Actually the stars are up there shining down on us all the time – day as well as night. The reason we can't see them in the daytime is because they are so faint (being so far away from us). We get blinded by all that bright light coming from the Sun and reflecting off the air and dust that make up our atmosphere.

It is only when the Sun sets that we begin to notice the stars. First we see one or two – the brightest of them. Then, as it gets darker, and our eyes get used to the dark, we begin to notice the fainter ones. I know it looks as though more and more of the stars are being 'switched on'. But that is not what is happening; they're 'on' all the time.

Why do stars twinkle?
Nisha

Light coming from the stars has to pass through the atmosphere to reach us here on the ground. The atmosphere can distort what we see. Have you ever noticed, for example, on a baking hot day if you look along a road everything in the distance tends to shimmer; it trembles and moves about. That is because the hot air close to the ground is lighter and rises, its place being taken by cooler air, which in its turn gets heated and rises, and so on. The light from the end of the road has to pass through moving pockets of air of different densities. This causes the light to get bent (in the same way as light bends when it passes from thin air into the dense glass in a pair of spectacles). That way the distant view is continually being distorted, as though you were looking at it through moving, crinkly glass.

The same thing happens when you look at a star. Its light is passing through moving air pockets of different densities high up in the atmosphere; that is what gives it the twinkle. But the star itself glows perfectly steadily.

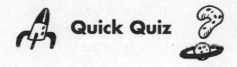

Quick Quiz

If you were an astronaut travelling to Mars, would the stars still seem to be twinkling?

 I was lying in bed counting sheep trying to go to sleep but I couldn't. So I opened the curtains and started counting stars. When I got to 396 it hit me. How many stars are there in the world?
Andrew Metcalfe (aged 9)

396? Well done! How did you make sure you didn't count the same ones twice? I wonder.

Anyway, as you probably guessed, there are a lot

more stars than that. On a clear night, with good eyesight, you might hope to be able to see about 6,000. But that's just a start. If you get hold of a telescope and you point it to a patch of sky where you think there are only a few stars, you will be amazed at what you can see: lots and lots of stars. The extra ones are so faint that you can't see them with the naked eye. Not enough light can get through the small 'window' in your eye – what we call the *pupil* of the eye (the black disc at the centre). But a telescope can collect a lot more light than that with its big lens or mirror. That's why the fainter stars can be seen through a telescope. And the bigger the telescope, the fainter the stars you can see, and the *more* stars you can see.

It is the same all over the sky. There are masses and masses of faint stars everywhere. It is especially so in the Milky Way. This is a faint spread-out band of light that stretches from one horizon, right across the sky overhead, to the opposite horizon. You can only see it on very dark clear nights and if you are away from street lights. If you live in a town, where there is a lot of light reflected from street lights, you might have difficulty noticing it at all. It is called 'The Milky Way' because that is what it looks like; it looks as though the milkman has had an accident and splashed very thin (probably skimmed) milk across the sky.

What actually is it? Why does it glow faintly? It is chock-a-block full of stars – so many stars you can't make them out separately. Their light just adds up to give a general glow.

So, back to your question: How many stars are there?

100,000 MILLION!

Phew! How are we to get our mind round a number *that* big? Well, suppose we decided to give each star a name. That's reasonable, isn't it? After all, our own star has a name: 'The Sun'. What it would mean is that every man, woman and child on Earth would have to think up twenty star names each – none of them being the same. That's how many stars there are!

They are all collected together in a great flattened disc – like a CD (only there is nothing compact about *this* disc). It is huge! You remember how we pictured the Earth (you on the sofa) one metre away from the Sun (the electric fire), and how, on that scale, the nearest star was 250 kilometres away. Well, I want you now to think of that 250 kilometres squashed down so that the nearest  star is only one metre from the Sun. On that new scale, the *furthest* star in the disc would be 25 kilometres away!

The disc is called 'The Galaxy'. Our Sun is about two-thirds of the way out from the centre of the Galaxy. When we look up at the Milky

Way, we are looking towards the centre of the Galaxy; we are looking at the disc of the Galaxy end-on. That's why there are so many stars in that direction.

So, there are a lot of stars. But I haven't finished yet. Read on . . .

39 How many galaxies are there?
Yasseen (aged 8)

I was talking of the Galaxy, with a capital 'G'. But you are quite right, Yasseen, there are more than one galaxy. Lots of them. It is difficult to know how many because the bigger the telescope, the fainter the galaxies we see as we probe further and further out into space. But it is thought that there are probably as many galaxies as there are stars in our Galaxy:

100,000 MILLION!

So now we discover that in order to give each *galaxy* a name, every man, woman and child on Earth would have to come up with twenty names – none of them being the same.

And yes, each of those galaxies has about 100,000 million stars. As I said, there are *lots* of stars.

 When there are new stars, what happens to the old ones?
Hannah (aged 11)

As a star becomes old, it gets to the point where it has just about used up all its fuel. Its nuclear fire is about to go out.

Now, I don't know about you, but I would have thought the star would then quietly fizzle out – like a camp fire dying down when it runs out of wood. But no! We're in for a surprise. In its old age, a star as massive as the Sun cools down all right; it becomes red-hot, instead of its usual white-hot. But, as I told Rebecca (question 12), it grows in size. Its fiery gas swells up. One day the Sun will become so large, it will almost swallow up the Earth. At this stage a star is called a *red giant*.

What happens then is that it sheds its outer layers, leaving behind at the centre a small white-hot glowing ball. This was the piping hot central core of the star. It is called a *white dwarf*. (Scientists come up with good names, don't you think?) The white dwarf then (at last) quietly fizzles out and becomes cold.

That's for a star like the Sun. Much more massive stars go out with a bang – and I mean that: BANG! One minute they look perfectly normal.

Then without any warning, there's a huge explosion. For several days, this one dying star is as bright as all the stars of the galaxy put together. What a way to go!

It's called a *supernova* explosion. As for what's left over after the explosion, we might find – close to where the centre of the star was – a *neutron star*. This is a dense ball about twenty kilometres across. When I say 'dense', I mean a speck of the stuff, the size of a grain of salt, would weigh the same as 10,000 crushed up lorries!

The neutron star gets its name, naturally enough, because it is made of neutrons. Remember (questions 1 and 2) how atoms are made of negatively charged electrons and a nucleus, and how the nucleus is made up of neutrons and positively charged protons. In the neutron star you can think of the electrons and protons being crushed together by the strong gravity forces, and then merging to form neutrons (their positive and negative electric charges cancelling each other out to give only neutral particles). These new neutrons then join the neutrons already there. So you end up with just neutrons.

And neutron stars spin. How they spin! Some whirl completely around several hundred times a second. We might not get dizzy on our spinning Earth; but we certainly wouldn't be able to walk straight after a ride on a neutron star!

That's one possibility for what you might find

where the old star used to be. Another is even stranger. Some old stars leave behind a *black hole*. I'm sure you've heard of those. What happens is that, while some of the stuff of the star is thrown out in the supernova explosion, most of it gets sucked down by its own gravity to a point smaller than the sharp end of a needle. From then on, anything passing close to this black hole is likely to get pulled in and squashed down. I shall have more to say about black holes later (questions 63–65).

 Why do shooting stars shoot? I hope you can answer it.
Lynette Hussey (aged 7)

Shooting stars. I love them. Did you know there are special nights for seeing them – times when there are likely to be more of them about? If the sky is clear, I always make a point of going out to have a look on 10 and 11 of December each year. On those nights, you can generally reckon on seeing them at the rate of about one a minute. You have to keep your eyes peeled. They shoot across the sky very, very quickly. They come without warning. Although they tend to point back to a particular position in the sky (the direction in which they are coming), you can never tell where exactly in the sky the streak will appear.

The important thing about shooting stars is that they are *not* stars. Although they appear to be about as bright as a star, they are not great balls of fire a long way off in space. Instead, they are small solid particles, generally no bigger than a grain of wheat. They are in fact bits of dust from comets. Having spent millions of years hurtling through space, they strike the Earth's atmosphere. In a matter of a second or two, they burn up, and disappear. The fiery trail makes up the shooting star. (Because it is not really a star, scientists prefer to call the trail by a different name: a *meteor*.)

Why do the particles burn? Friction. You know how on a cold day you can warm your hands by rubbing them together. We call it heating by *friction*. Well, when the particle enters the atmosphere, about a hundred kilometres above our heads, it is travelling fast – tens of kilometres per second. As it rubs against the air, it heats up to the point where it glows brightly, melts and boils away. And that's the end of it.

If the particle is bigger, its core might manage to reach the Earth's surface before it has all boiled away. As I mentioned in the answer to question 31, we call such rocks from space *meteorites*. Remember, meteorites are what caused the craters on the Moon.

Why are some nights better than others to go looking for shooting stars? I told Quentin (question 21) that comets, when they get near the Sun, give off dusty clouds – the comet's tail. If the Earth, on its own journey round the Sun, passes through what's left of that tail, then the dust particles hit the atmosphere and form shooting stars. One year later, when the Earth arrives back at the same place in its orbit, it passes through the same cloud of dust again. That's why the display of shooting stars is then set to repeat itself at the same time of the year as before.

So, I expect you out in the backyard on 10 and 11 of December – but remember to wrap up warm.

Hint: I do my shooting star watch lying out flat in a deckchair. That way I don't get a crick in the neck from looking up. What I say is: there's no point getting old, if you don't get clever with it!

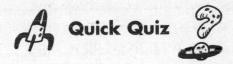

Quick Quiz

(a) If shooting stars are just tiny grains of sand burning up, how come they look as bright in our sky as proper stars?
(b) Would an astronaut on a space journey get a better view of shooting stars?

The Universe

 How did the story get around about the Big Bang and who told the story? Was it a girl, boy, man, or woman?
Krystle Lakee (aged 11)

When discussing the galaxies (question 39), there was one thing I forgot to say about them: they are all rushing away from us. The American astronomer, Edwin Hubble was the first to notice that the further away a galaxy was, the faster it was hurtling off into the distance. Carry on like this and they'll all be gone!

Now we mustn't take this personally. There is nothing specially wrong with us that all the other galaxies want to get as far away from us as they can. It wouldn't matter which galaxy we were on, it would appear that all the others were going away from us. That is because the distances between *all* galaxies is getting bigger and bigger. But why?

The simplest answer is to imagine everything in the Universe starting out squashed up together. It then suddenly EXPLODES. All the bits fly apart, getting further and further away from each other –

which is exactly what we see the galaxies doing today. The explosion is called 'The Big Bang'.

But how can we be *sure* there was a Big Bang? If we are right, then the Bang must have been very violent. Violent explosions are hot; they give out a lot of heat and light – as happens when a bomb goes off. The flash of light from the Big Bang should still be around in the Universe today – somewhere (there's no other place for it to go!). All right, it will have cooled down a long time ago; you wouldn't expect to be able to see it with your eyes any more. These days it would be more like radio waves, or the waves you might use in a microwave oven. You can't see those either. But with the right equipment (a Walkman?) you can detect invisible radio waves.

Two other American scientists, Arnold Penzias and Robert Wilson, became the first to discover the radiation from the Big Bang. It is coming down to us out of the sky all the time, day and night. It causes interference on television. If your TV set isn't tuned in properly to the station, you get inter-ference – the picture looks like a snowstorm. Well, about one out of every hundred of those 'snowflakes' is due to the flash of light from the Big Bang. Quite a thought, eh? You didn't know you had a Big Bang detector right in your home, did you!

So that's how the idea of the Big Bang got around.

What was there before the Big Bang? It couldn't have been nothing.
Sophia Dabnor (aged 10)

We haven't a clue. Why? Because the density of matter in the early Universe was so enormous. Today, if we could smear out all the matter to fill up the whole of space, it wouldn't amount to anything more than a very, very thin gas. Now imagine going back in time, further and further – towards the moment of the Big Bang. The gas gets thicker and denser. It becomes as dense as water, then as dense as rock, then as dense as lead, and so on. In the end, when we get right back, all the way to the very moment of the Big Bang, the density would become mega colossal – it would become infinite. (Infinity is the biggest number anyone can ever think of, multiplied by umpteen zillions, and then you add on a few squillions for luck.) Infinite density is something we scientists cannot handle. And when you grow up and it's your turn to be the scientist, I don't see how you, or anyone else, is ever going to get round this problem.

In fact, when you get to question 66, you will discover a second reason why there is probably no answer to your question.

How old is the Universe? Please try to answer back because I have been wondering for ages.
Rosie Bunker (aged 9)

The answer is about 12,000 million years!

'Now, how can he possibly know that?! Where are the witnesses?' I hear you ask. No, there were no witnesses. Human beings came on the scene for the first time only a few million years ago. (It's funny how in this sort of work you get to thinking that a few million years is *'only'* a few million years.) So, how do we know the age of the Universe?

As I have just been saying, the galaxies today are still flying apart because of the Big Bang. The further away they are, the faster they are going. If a galaxy is five times further away than another, it will be travelling five times as fast; twenty times further away, twenty times as fast; etc. Now, you don't have to be a genius to work out from this that if we imagine going *back* in time, a point will be reached when all the galaxies were together at the same place. Not only that, knowing how fast the galaxies are moving, and how far they have to go to get on top of each other, you can work out *how long all this takes*. That in turn tells you *when* the Big Bang happened, ie. 12,000 million years ago. So, that's how we work out the age of the Universe – assuming it was created at the moment of the Big Bang.

Quick Quiz

Two cyclists decide to race each other. They set out at the same time. One travels at a steady speed of twenty miles per hour, the other can manage only nineteen miles per hour. The race is stopped when the faster one has covered a distance of sixty miles, the other a distance of fifty-seven miles. How long did the race last?

45 Where is the centre of the Universe?
Christopher Moore (aged 10)

It's a good question. You'd think that with the Universe beginning with a Big Bang, the explosion must have taken place somewhere. There ought to be a spot you can visit that has a notice saying 'The Big Bang took place here in the year 12,000 million BC.' Alongside there ought to be a souvenir shop – and a café.

In fact, the Big Bang is much more interesting and mysterious than that. It was no ordinary explosion – not the sort where a bomb goes off at one particular point, throwing its debris out into the surrounding space. With *that* kind of explosion you *can* pin-point where the bomb must have been originally.

The Big Bang was a special kind of explosion – the only one of its kind. At the time of the Big Bang, not only was all of the matter of the Universe squashed down to a point, *all of space was squashed down to a point too*. There was no space outside the Big Bang. This means we have to think of the Big Bang happening *everywhere*. It is because the Big Bang happened absolutely everywhere, when space was squashed very tiny, you can't think of a particular point being the centre of the Universe – a point where the Big Bang happened and from which all the galaxies are now rushing away.

Just like you, I need help when trying to wrap my mind round these difficult ideas. What I do is this: I imagine a balloon. It starts off very tiny.

But then, as it is blown up, it gets bigger

. . . and bigger.

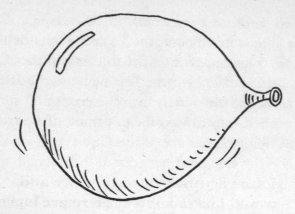

If two flies land on the balloon, they find that they get further and further apart. It is not that they are walking away from each other; they move apart because the rubber in between them is expanding.

It's the same with the galaxies; the distance between them increases, not because they are moving through space, but because the space between them grows, and as it grows, it carries the galaxies along with it. Space continues to grow today, as it has done ever since the moment of the Big Bang.

Just as you can't look at the surface of the balloon and pick out one particular point from which all the rubber expanded ('the centre of the balloon's surface'), so you can't look at the Universe and pick out any special point where all of space started to expand ('the centre of the Universe').

I am very interested in space and would like to know if there are lots of other universes, or is there only our universe?
Fatima Meho (aged 11)

It's difficult to say. For a start, it depends on what you mean by the word 'universe'.

With more and more powerful telescopes, we can see fainter and fainter galaxies; this means we can look further and deeper into space. But there is a limit to how far we can look. The reason has to do with the fact that it takes time for light to travel from one place to another. It takes eight minutes for light to get from the Sun to Earth; it takes four years to get from the nearest star. What is the maximum time light can take to reach us? Answer: the age of the Universe. Obviously it couldn't have been travelling for longer than that. So, we cannot expect to receive light today that has been travelling for longer than 12,000 million years. That in its turn means that it could not have come from a distance greater than that which light can cover during such

75

a time. We cannot know of the existence of anything that lies further than that. Everything that is *within* that distance, and from which we *could* receive light, is said to be part of 'the observable Universe'.

But we think more than *that* exists. Why? Well, as time goes by, we receive light from further and further away; this is because it has had a longer time to travel. The observable Universe takes in more and more of what lies *outside* it. So, we can say that as well as our observable Universe, there is the rest of the Universe outside – further off into space.

But that still leaves the question of whether there are universes that have their *own* spaces and times – spaces and times that are not part of ours. Indeed, they might have no space or time at all! It's hard, if not impossible, to imagine such a thing. But who knows? There might be universes built on quite different lines to ours. We just don't know. And what's more, we shall *never* know. If they are not part of our time and space it is impossible to see how we could ever get any messages from them – how we could ever make contact with them. And without that, we can never get proof that they exist.

How will the world end?
Christopher Moore (aged 10)

We can't be sure, but here are the possibilities:

All the galaxies are flying apart because of the Big Bang. But every galaxy is pulling on every other galaxy with its gravity. So that means the galaxies are gradually slowing. If they keep this up, they will eventually come to a halt; the expansion of the Universe will be over.

What will happen then? Well, the galaxies can't just sit there in space doing nothing. The gravity between them is still pulling on them. So, from now on, the galaxies start *coming together*. Instead of expanding, the Universe now starts to shrink. This carries on until all the galaxies come piling in on top of each other . . .

KERPOW!

We call this a *Big Crunch*. The end of the Universe.

So is that, in fact, how the Universe is going to end? Possibly. It depends on how strong the gravity forces are. After all, as the Universe expands and the galaxies get further apart, the force between them gets weaker. It could be that the force vanishes more or less to nothing before it has managed to stop the galaxies in their tracks; the galaxies will then be able to get clean away, and the expansion will go on for ever. This is the other possibility.

What will happen then? The stars will gradually use up all their fuel; their fires will go out. As I was telling Hannah (question 40), some stars end up as burnt-out cinders, others as cold neutron stars; others as black holes. Life on all planets will come to an end. And that will be that. We call it the *Heat*

Death of the Universe.

Shiver . . .
Shiver . . .
Shiver . . .
Shiver . . .

OK, what's it to be: Big Crunch or Heat Death? Frankly, we don't know. It all depends on the strength of the gravity force. Is it, or is it not, powerful enough to stop the expansion? To answer that, we must know the average density of matter in the Universe (the average amount of matter in a given volume – in a cubic metre, say). This is because it is matter that causes gravity. Obviously, the more stuff there is in the Universe, the more gravity. So, how much stuff *is* there?

There's no problem adding up the matter in the stars – that's easy. The trouble is: what about all the stuff out there that we *can't* see – because it isn't glowing? We call it *dark matter* – for obvious reasons. And there seems to be an awful lot of it; probably a hundred times as much as the stuff we can see. What's more, we aren't really sure what the dark matter is made of. Is it like the other stuff, only it's not shining, or is it a kind of matter we haven't come across before? All of which is very embarrassing for us scientists. Here we are, the great experts on the Universe, and we haven't a clue what ninety-nine per cent of it is?!

But one thing we do know. The density is very close to what we call the *critical value*. That is the value of the density you would get if the Universe was teetering on the brink between expanding for ever, or ending up in a Big Crunch. Now that is odd – very, very odd. Of all the values it could have been, why is it so close to the critical value? A co-incidence? Unlikely. It seems to us that there must be some special reason why it *must* be spot on the critical value. In fact, we now have an idea as to what might have happened in the early stages of the Big Bang to force it to have this special value. If we are right, then that means the Universe will keep on expanding for ever, gradually coming to a halt – but only in the infinite future.

Whether this is so or not is one of the first things boys and girls in school today have to find out when they become the scientists of the future. But whichever way it turns out, the important thing to remember is that you should not lose any sleep over it. The end of the world is not going to happen for a very . . . very . . . very long time.

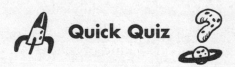 **Quick Quiz**

If we are right, and the density really does have the critical value, what will be the end of the Universe: Big Crunch or Heat Death?

What is there when you go through clouds and you go through space and then you go out of space? What comes up next?
Joseph (aged 5)

What we think happens is this:

You go up through the clouds. Then you go up through the rest of the air – higher than any aeroplane. Then you are out in space among the Sun, the planets and the spacecraft. Then you leave them behind, and start going past the stars (which, like the Sun, are great balls of fire). Then you go past more stars – and more stars – and more stars – and more stars – and more stars – and more stars – and more stars – and more stars – and more stars – and more stars – and more stars – and more stars – and more stars – and more stars – and more stars . . . .
. . . .
. . . .
. . . .
. . . .
. . . .
. . . .
. . . .
. . . .
. . and so on, until you have filled up all the paper in the world writing 'and more stars'. Even then you have only just begun your journey.

In fact, you never come to the edge of space. Why?

Because if you ever came to the end of space,

what would there be after that? Nothing – obviously. But nothing is what space looks like! So it would be *more* space, and you would *not* have come to the end of it. That's why we think space goes on for ever.

However, we *might* be wrong! It might be that if we were to go on such a journey, after a certain time we would find ourselves back where we started here on Earth! How can that possibly happen?

Well, suppose instead of a space journey you went on an ordinary aeroplane trip. Imagine you are flying due east and you kept going and going in that same direction – always flying due east. What would happen? Would you get further and further from your starting point?

At first, yes. But after a while, when you are on the other side of the Earth, you would not. While still flying due east, you would eventually find your-self back where you started. It's all because the

Earth is not flat (as it looks), but is really a great big round ball. Without knowing it, you were flying around a large circle, even though the navigator's compass kept pointing in the 'same' direction.

Now, we don't know, but it might be like that with space travel. Space might be curved in some very peculiar way so that, although we think we are always travelling in the same direction (a straight line out into space) we actually end up back where we started. That might be another way of never coming to an edge of space.

One piece of advice: if you ever decide to become an astronaut, think twice before agreeing to go on a journey to find out which of these answers is the right one. It could be a pretty long and boring trip.

What is space made out of?
Farah Morris

Some people think 'Space is simply nothing – and
that's all there is to it'. But actually your question is
a very good one. The answer is not obvious.

To see why, let me ask you a question about air.
How do you know that, at this very moment, you
are surrounded by air? 'Easy,' you might say. You
show me by blowing out your cheeks. The cheeks
stay like that because of the squashed up air in your
mouth. Or you might wave a sheet of paper about,
setting up a wind. But note what you are doing in
both cases; you are *disturbing* the air. The squashed
air in your mouth is denser than that outside. As for
the sheet of paper, when it is waved, there is more
air piled up in front of it than behind.

But suppose you sat perfectly still, and didn't
breathe. With the air undisturbed – the same everywhere, it would be much more difficult to know it
was there.

Now I want you to imagine a kind of 'air' that is
exactly the same everywhere. And I mean EVERY-
WHERE – it fills the whole of space. (By the way,
start breathing again!) It's a kind of 'air' that, as you
move, it slips through you without you noticing it; a
kind of 'air' where there is no pile up of it in front of
you, or hole in it left behind; a kind of 'air' where
there is always exactly the same amount of it
whether you are thinking of a region of space outside

your body, inside it, or in the far depths of space.

Such a completely undisturbed 'air' would be impossible to detect; it would be invisible. Why invisible? Well, think for a moment about this book in front of you. How is it you are able to see it? Because you can look at it from the outside. You can say 'It's here right in front of me on my lap.' But with the kind of 'air' we are talking about, it's not like that; it is everywhere; it has no edge; you can't see it from the outside because you can never get outside it. That's why it is invisible.

Now the amazing thing is that this is how we scientists sometimes think of space – so-called 'empty space'. We don't think of it as *nothing*; we think of it as a kind of 'air' that is the same everywhere. And because it is the same everywhere, we cannot see it and it looks and behaves – like nothing!

'How stupid!' I hear you say. But hold on. I haven't finished yet. If this space-like 'air' *always* remained the same everywhere, it would be undetectable. If that were the case, I agree the idea *would* be stupid. But it isn't so. In very special experiments, scientists *are* able to disturb this space-like 'air'. We can actually knock holes in it! When we bash space really hard, we can knock a bit out of it; we see the particle we've knocked out (which used to be part of the space-like 'air'), and we see the hole left behind (which also behaves like a particle – we call it an *antiparticle*).

Who could have thought that *nothing* could be so interesting!

Gravity

Why is the world a sphere and not a pyramid, a cylinder, a cuboid or a tri-angular prism?
Natalie Moon

Everything pulls on everything else. If you have two objects placed in space, it doesn't matter what they are (two atoms, two elephants, you and me) each pulls on the other; they try to come together. As we learned from question 9, this pulling force is called gravity. The more massive an object, the stronger its gravity force. And the closer the two objects, the stronger the force becomes.

The Earth originally formed out of a cloud of dust and gas. Each particle of the cloud pulled on every other particle. They got closer and closer together, and the gravity forces got stronger and stronger. The bits of dust and stuff pressed into each other and tried to pack as close together as they possibly could. And the way to pack things tightly so that they are as close to each other as they can get, is to have them form a round ball; it is simply the neatest way of packing things together. That is why the Earth, and the planets, and the Sun and Moon all ended up round.

Things roll down hill, so why don't they roll up?
Sophie Hibbs (aged 10)

It's all due to this gravity force I was telling Natalie about. Things on a hill are like the bits of dust that originally came together to form the Earth. Just as all those bits tried to get as close together as they could, so the thing on the hill behaves the same way; it too is trying to get as close in as it can to all the rest. OK, the stuff that makes up the *top* of the hill is pulling it in the opposite direction: upwards. But all the rest of the Earth is pulling it down. So it is little wonder that it is the rest of the Earth that wins, and the thing rolls down rather than up.

Which is a pity. I could save myself a lot of money on petrol if I could persuade my car to roll up hills as well as down.

 If you were standing at the South Pole, why aren't you upside down?
Angela Chahwan (aged 9)

All of us start out by thinking that the Earth is flat and horizontal, and that there is a special direction in space called 'down'. That is the direction everything gets pulled. Jump up in the air, and you get pulled *down* again on to the Earth's surface. As you sit reading this book, you are being pulled *down* into your chair.

But that isn't the best way to think about it. As we have seen, the Earth is not flat; it's a round ball.

The Earth's gravity tries to pull you towards the centre of the ball. And not just you. It pulls on everyone and everything, wherever they are – at the North Pole, the South Pole, the equator, or in between. They all get pulled towards the Earth's centre.

Everyone says they are being pulled *down*. By the word 'down' they mean the direction to the Earth's centre from wherever they happen to be on its surface. The trouble is, all their 'downs' are different! If you are at the North Pole, your 'down' is in the opposite direction to that of someone who is at the South Pole.

It's a good thing there isn't just the one 'down' direction. If there were, and it was *our* 'down', the people living on the opposite side to us would have to hang on for dear life. Not only that . . .

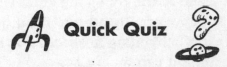

Quick Quiz

. . . what would happen to all the air, and the water of the seas and oceans as they got pulled round to the other side?

How do the planets and the Sun stay in the sky and why don't they fall?
Leonie Lambert

I have said how everything pulls on everything else with its force of gravity. And this force stretches right across space. It gets weaker the further away things are, but it is always there; it never goes to nothing. That means the Earth and the Sun are pulling on each other – and the Moon – and the planets. They are all pulling on each other. So, you're quite right; you would expect them all to fall in on top of each other as one big heap. But for some reason they don't. Why?

The answer is that they are all going round each other. The Earth and the planets go round the Sun; the Moon goes round the Earth. How does that help? Well, suppose you tie a brick on to the end of a piece of rope, and then whirl the brick around your head. (But DON'T do this near windows!) You are pulling on the brick, but the brick doesn't get any closer to you. All your effort is needed just to keep the brick going in a circle. If you stop pulling (i.e. let go of the rope) you know what happens – the brick goes flying off into the distance.

That's what is happening in space. The Earth is pulling as hard as it can on the Moon, but it doesn't fall out of the sky because it is going in orbit around us. In the same way the Earth and the other planets don't fall into the Sun and get burned up in its fires because we are going in orbit around the Sun. We are all running around in circles – and it's lucky for us that we are.

My question is how does the Earth keep going around the Sun?
Rachel Capeless (aged 9)

When riding your bike, you know how you have to keep pedalling to keep going. In the same way, when your mum takes her foot off the accelerator, her car slows down. Or suppose you send this book sliding across a table top, it soon comes to a halt. From these examples it's obvious that moving things tend to stop. In order to keep something moving, you have to keep pushing it. So, if the Earth carries on travelling at high speed around the Sun, something must be pushing it – right?

WRONG! One of the things I love about science is the way it often catches you out. What at first sight appears obvious, turns out to be quite the opposite. And that is how it is with motion. To keep something moving at a steady speed you do *not* have to keep pushing it . . .

'Huh!' I hear you grunt. 'What about the bike, the car and the sliding book?'

Allow me to finish. I was saying that you do not need to keep pushing – *assuming there aren't any other forces trying to slow it down*. That's the important bit that often gets overlooked; it's the other forces that make all the difference.

As the book slides over the table there is friction; the surface is rough to some extent. And this slows the book down. The same is true of the car and the bike. You get frictional forces wherever one surface rubs over another – where the centre part of the spinning wheel comes into contact with the bike frame, for instance. You can reduce this slowing down force to some extent by using ball bearings, but you can't get rid of it completely.

Then there is air resistance. As you and the bike move forward, you have to push the air out of the way to make room for yourself, and that takes effort. Again you can reduce this (by crouching low over the handlebars), but you can't get rid of this slowing-down force either – not completely. So, that's why you need to keep pushing on the pedals; it is to overcome the slowing-down forces.

What has this to do with your question about the Earth's motion? Simply this: out in space there is no air, so there is no air resistance. Not only that, the Earth is not rubbing against any surface, so there is no frictional force. In fact there is nothing trying to slow it down. That in turn means we don't have to keep pushing on it to keep it on the move. Which is just as well; imagine what kind of engine you would need to keep a 'vehicle' as massive as the Earth going – and just think of the fuel bill!

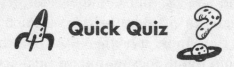 **Quick Quiz**

Why do you think the hull of a ship is shaped the way it is?

I would like to know how the tide goes out and in.
Nicholas Bae (aged 7)

The gravity of the Moon pulls not only on the solid Earth but also on the seas and oceans. Water, being 'loose', is easily distorted – it gets pulled out into an oval, rugby ball shape. The solid Earth, on the other hand, remains roughly round.

As the Earth spins on its axis, the bulges in the oceans tend to stay pointing in the same direction, towards the Moon. That means twice in every twenty-four hour day, a point on a coastline will find the water gets higher, then lower. That's what we call the *tide*.

Actually it's a bit more complicated than that because the Sun also pulls on the water, trying to make it oval-shaped. Because it is so much further away than the Moon, it has less than half the effect of the Moon, but it does produce a tidal action of its own. When the Moon and Sun are lined up with the Earth, all three along the same direction (twice

91

a month), then the two tidal effects add up and one gets especially high and low tides. When the Sun and Moon are pulling at right angles to each other, then the difference between high and low tide is not as great.

5 6 How do you float in space?
Philip Browning

The way astronauts float around in space is very confusing. If you are standing on the Earth, and you jump up, you quickly get pulled down to the ground again by gravity. You don't float. So, does that mean there is no gravity where the astronauts are?

No. As I was telling Leonie just now (question 53), Earth's gravity stretches right out into space. The astronauts are being pulled towards the centre of the Earth just like we are. So, the fact that they float can't be because there is no gravity where they are. The answer is that, with its rocket engines switched off, the astronauts' spacecraft behaves like the Moon; it just coasts along in orbit round the Earth. The Earth's gravity is still pulling on it, but the 'pull' is all used up in keeping the spacecraft going roughly in a circle; there is none left over to bring it down onto the Earth's surface. The space-craft has, in fact, become an extra mini 'moon' of the Earth.

The same is true of the astronauts. They are themselves coasting along, cancelling out the gravity pull on themselves by going in orbit around the

Earth. They have become 'mini-moons'! But their orbit is the same as that of the spacecraft. So they coast along together. That's why, when we see them inside the spacecraft, or going for a space walk alongside the craft, they appear to be floating. It's a bit like two cars travelling along the motorway side by side at the same speed. To the passengers, as they look across at each other, they don't seem to be moving, but in fact, they are both speeding along.

57 Why is there no air in space?
William Goock

It's due to gravity – again. The Earth, the Sun and the planets all pull on the air and the other gases in space, just as they pull on everything else. So that's why you find the air hugging the Earth's surface and not spread out evenly through space.

Because of this, you might expect all the air to end up flat on the ground – in the same way as rain-drops get pulled down by gravity and end up as puddles. (If that were so, we would have to spend our lives crawling around on hands and knees with our noses on the ground sniffing up the air!). It's not like that because air is a *gas*. And one of the things about gases is that their smallest parts (the molecules of the gas) are always rushing about like mad; they can't sit still for a minute (like some people I know). If there were no gravity to hold on to them, they would fly off into space like dogs let off the leash. That's why the air doesn't simply lie around on the surface of the Earth, but spreads upwards a little. The molecules are always trying to escape off into space, but they keep getting pulled back before they have gone far.

The air is thicker near the ground, and then thins out higher up. As you probably know, when mountaineers climb really high mountains they often take their own supply of oxygen with them and wear oxygen masks. If they didn't, they would be panting all the time and have to keep stopping to catch their breath. Modern jet planes have the same problem. They fly so high that air has to be pumped up into the cabin so passengers and crew have enough to breathe.

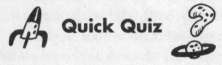 **Quick Quiz**

Why do you think the Moon doesn't have an atmosphere?

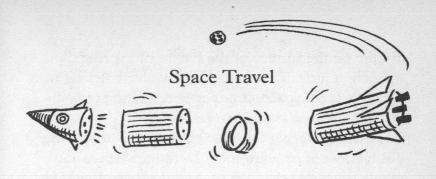

Space Travel

 Why do rockets have lots of bits if there is only one bit that comes back to Earth?
Faye Watkins (aged 7)

A typical rocket launch uses up a colossal amount of fuel. So the rocket must be equipped with very large tanks for containing it. These are heavy and it takes a lot of fuel just to lift the tanks themselves. Once a tank is emptied, it is no longer needed; it has done its job. So why use up even more precious fuel lifting it any further? The most economical move is to get rid of it now. And that is what happens. The empty tank – together with engines and any other parts that have done their job and are no longer needed – get detached and fall back to Earth, leaving just the space capsule, or the space shuttle to make the final journey. It is the simplest and cheapest way of making these space journeys.

NOTE! The fact that space engineers behave like this is *not* be taken as an excuse for throwing empty Coke cans out of the car window just because they are empty and no longer needed. People who leave messes for others to clear up make my blood boil!

59 Why is a rocket shaped how it is?
Amy Hickson (aged 9)

I was telling Rachel (question 54) about the slowing down effect of air resistance. To reduce this to a minimum, I said that she ought to crouch down low when riding her bike. The same is true when designing a car or the Eurostar express train. They are made smooth and sleek so they can slip more easily through the air. That way they do not need to be pushed so hard to keep them going at a steady speed, and their engines can do more miles per gallon of fuel.

Because a rocket starts its journey by having to go up through the Earth's atmosphere, it too has to keep air resistance to a minimum. That's why it is designed to be smooth and pointed.

But when it comes to the capsule – the bit that makes the final journey across space – it can be more or less any shape because there is no air out there to worry about. That's why the Apollo craft that landed on the Moon didn't look anything like a rocket.

Mind you, if you have a craft that is to return to the Earth after its mission, it has to complete its journey by passing through the Earth's atmosphere once more. But now you are not worried about reducing the slowing down force of air resistance. Quite the reverse. The craft is travelling at high speed and has to end up with no speed at all by the time it lands on the Earth's surface. This time the aim is not to reduce air resistance, but actually make use of it. The craft needs to have a wide front that will *increase* the slowing down force of air resistance. This front part of the craft is called the 'heat shield'. As the craft enters the atmosphere, there are strong frictional forces as the air 'rubs' over the shield's surface, and these heat up the shield to glowing hot temperatures. That is how the craft is able to slow down and land safely. It just goes to show that air resistance is not always a bad thing.

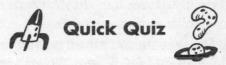

Quick Quiz

When the spacecraft is speeding through space it has lots of energy. When it lands gently, all this energy of motion has gone. What has happened to this energy?

Why do astronauts wear space suits?
Marco (aged 8)

Like everyone else, astronauts need to breathe – of course. Air contains a gas called 'oxygen'. We have to keep on breathing in oxygen to stay alive. But, as William was pointing out just now (question 57), there is no air in outer space.

Astronauts have to take their own supply with them. The cabin of their craft is provided with oxygen, so they can breathe normally. But if they go outside the craft, to take a space walk, then they have to put on specially sealed spacesuits and helmets. Containers are strapped to their backs, and these feed the precious oxygen into them.

61 Can astronauts cry in their high pressure spacesuits?
Anon

Although spacesuits are bulky, the pressure inside them is not particularly high. In fact, the aim is to have a pressure similar to what we are used to here on the surface of the Earth. They have to be strong and tough to make sure that they don't get easily punctured, so allowing the air to escape out into space where the pressure is zero. So, if you were wondering whether the tears wouldn't come out of the astronaut's eyes because the pressure would keep them in, then that is not the case.

However, he (or is it she?) would still have a problem. The eyes would fill up with tears, but they would not run down the cheek; gravity would not be pulling them down. (Perhaps that is what you had in mind in asking the question.) So the astronaut would have difficulty seeing. For that reason it would be a good idea to cheer up as quickly as possible.

Actually, if I were an astronaut, crying would not be my biggest worry. What I remember most about that wonderful film, *2001* – all about space travel – was the man who wanted to go to the toilet and had first to read a long list of instructions as to how to do it under conditions of zero gravity!

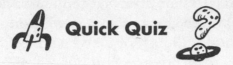

Quick Quiz

When astronauts go for a space 'walk', they aren't actually walking. Why?

62 How many aliens are there in outer space?
Katie

This is a question I am always being asked. I'm afraid I don't know the answer. Nobody knows whether there are space aliens or whether we humans are the only form of intelligent life in the Universe.

Searches are being made. Scientists are listening for radio signals which might have been broadcast to us by aliens from another planet in space. But nothing has been found yet.

What are the chances of there being life out there? I think they're pretty good. There does not seem to be any on the other eight planets going around the Sun. But there are lots of other suns. As we saw in questions 35, 38 and 39, each star in the sky is a sun, and there are billions and billions of them.

We expect many of them to have planets going round them. In fact, the first planets belonging to different suns have now been discovered – about ten of them so far. Of course, like most of the planets belonging to our Sun, they will either be close in to their star and too hot for life to develop (like our planet Mercury), or their orbit will take them too far away from their star and they will be too cold (like our planet Pluto), or they might be too small to hang on to an atmosphere. For any of these reasons, most planets will have no life on them.

But it is also expected that, by chance, there will be planets at just the right distance to have a nice warm temperature and enough gravity to be able to hold on to an atmosphere. They might also have water. Given such a planet, there is at least a chance of life starting up. After all, we know that in the clouds of dust out of which stars form there is to be found the very kinds of material that go to make up our own bodies.

The first forms of life would be very simple to begin with – tiny, tiny bugs. But with time these could get bigger and more interesting. Who knows, in the end they might finish up like us. They wouldn't *look* like us humans – that would be *very* unlikely – but they could be as intelligent as us. They could broadcast to us. They might even visit us! But before we get excited about that, we must remember that any space aliens would live a very, very, very, very long way from us. It would be extremely difficult for them to travel so far. It's not just a matter of saving up for the fare, the journey would take so long, they would be dead before they got here – which makes it all a bit pointless, don't you think?

'But,' you might be saying, 'what of UFOs and those funny circular shaped patterns in corn fields? Aren't they due to aliens coming from space?'

Well, all I can say is that you'll have to make up your own mind on that score. For myself, I think the corn circles were someone's idea of a joke. As for the UFOs, some of them are known to be hoaxes, some are known to have been harmless things like weather balloons floating in the sky. As for the rest, who knows?

One thing I think you have to bear in mind is that the distances to planets belonging to different suns is so great it would stretch space technology to the limit – and frankly I don't think the aliens would bother to come. After all, if their technology is *that* good, they can probably pick up our radio and TV transmissions. Now, I ask you: do you honestly think that an alien tuning into one of our typical chat shows is going to think 'Gosh, what wise, clever and interesting people. It would be worth travelling 40 million million kilometres to have an intelligent conversation like that'?

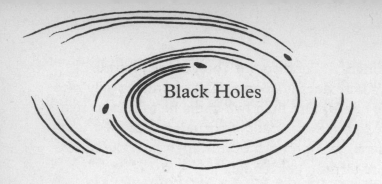

Black Holes

 What is the black hole in space?
Daniel Bilton

Black holes give me the shivers. They are really, really nasty. What happens is this:

As we have seen, each star is a great ball of fiery gas, just like the Sun. Because it is so hot, the gas jiggles about like crazy. It's only the strong gravity of all that gas that keeps it together in a round ball.

But, of course, as I was telling Hannah (question 40), a star can't burn for ever; like any other fire it begins to run out of fuel. The supply of heat is being cut off, but the gas is still shining and losing heat to space – so it cools down. The jiggling gets less. This means gravity can get a tighter grip on the gas and pull it in closer. But the closer the atoms of the gas get to each other, the stronger becomes their gravity force, so they get pulled tighter still, which means gravity gets even stronger, which means the gas gets packed even tighter, which means the gravity gets even, even stronger, which means the atoms get even, even more tightly packed, which means gravity gets even, even, even stronger . . . and so on. You get the picture.

Where does it all end? If the star is a really massive one – more than two and a half times as massive as the Sun – it all of a sudden collapses down. There is a supernova explosion, and at the centre there forms a black hole.

The gravity of a black hole is so enormous that anything getting too close to it gets sucked in. And once you have been sucked into a black hole, that's IT – you've had it, you'll never be able to get out. In fact, gravity is so strong near a black hole even light can't escape its grip – that too gets sucked in.

Everything that falls into a black hole ends up at a point at its very centre. I don't know about you, but I find it absolutely amazing that a huge thing like a star (a million planets the size of the Earth would fit into the Sun) gets crushed down to something smaller than the point of the finest needle.

Nor does it stop there. We believe that black holes can form in a second way. Recall how the stars are gathered together in galaxies (question 38). All the stars in a galaxy pull on each other with their gravity forces. The reason why they don't all end up at the middle in a heap is that they are in orbit around the centre of the galaxy; the galaxy as a whole is rotating like a whirlpool.

At least, that is what the stars normally do. But sometimes a star will come particularly close to another one and have its path deflected. If it's unlucky, it could now find itself heading for the centre of the galaxy. In time lots of stars are likely to end up at the centre. With nothing to keep them apart, they crash into each other and fuse together to form a really massive black hole – one that might contain millions and millions of stars. It is thought that most galaxies have a massive black hole at their centre. The more stars the hole gathers up, the stronger becomes its gravity force and the more likely it is to capture yet more stars. It could be that all the stars of a galaxy eventually get sucked in!

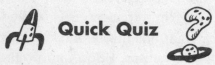

Quick Quiz

Why is a black hole black?

The two of us are very interested in science. If it is possible we would like to know a lot more about black holes in space. We both have talked about it for quite a few months now. We understand that nobody has tried to enter one because they think it is a one-way tunnel. We both think that there is something on the other side. We know the consequences are very dangerous, but we're both curious about it and we would love to have permission to enter it. Could you possibly send us a few addresses of people

**who could tell us more about fulfil-
ling our dreams.**
*Names withheld. (The writers are
two boys serving sentences at the
Young Offenders' Institution,
Portland, Dorset.)*

I am not quite sure why you are so interested in
whether you can go into a black hole and come out
the other end. I don't know much about the kind of
place you are in at present, but do I take it that it is
the kind of place people try to escape from? If so,
then your idea of using a black hole as an escape
route is very ingenious. I am sure it has never been
thought of before. If you did find a black hole handy
and you went into it, you can be pretty sure no one
will come looking for you in there. That is the good
news. The bad news is that you would be dead –
and I don't mean just 'dead' – I mean SERIOUSLY
DEAD!

If you were to go in feet first, then your feet
would be ripped off your ankles as they got sucked
in first; then the rest of you would follow with your
ribs being crushed in from the sides, and finally the
whole of your body would be crushed down to a
point. It would be like the way old cars get crushed
down to a small block of scrap metal – only you
would be so small, no-one would be able to see you.
It would be over very quickly – but boy, it would
hurt.

What would happen to you next? Most scientists think that that would be the end of you – you would remain squashed at a point. But there is an idea floating around that that might not be true. Having been squashed down to a point, you then travel along a tunnel called a 'wormhole' and your bits get squirted out of a *white* hole in another universe. Freedom at last; you have made your escape! Except that you come out like soup that has been through a blender.

No, I suggest you knuckle down, do your sentence, learn your lesson – and then get jobs which allow you to make use of your interest in science.

 How come we know about black holes if you do not come out of them? I want to be a scientist – because it would be exciting.
James Ockenden

That's a very good question. Not many people think about that. But you are quite right: no one can come back out of a black hole to tell us what it is like in there. Not only that, black holes don't give out any light of their own (like a star) and don't reflect any light that shines on them (like the Moon). So we can never hope even to see a black hole, let alone visit inside one. So, why are we scientists so cocky about them being there?

Let me begin by reminding you of the story of the Invisible Man. There was this man who made

106

himself invisible; you couldn't see him at all; you looked straight through him. When he wanted to show people where he was, he would wear clothes and would wrap bandages round his head. That way, by watching what these clothes were doing, people could 'see' that he was there in the room with them. Except, of course, they didn't actually see *him* ; it was the clothes and bandages they saw. (If you get a chance of viewing *The Invisible Man* film on TV, don't miss it. It's in black-and-white, but very good – most black-and-white ones are.)

It's a bit like that with black holes. You can't see the hole itself, but you can see what it's doing to the things close by. For a start, many stars come in pairs. Instead of having planets going around the Sun, you get two suns (meaning stars) going round each other. Sometimes one of the stars, when it's old and burnt out, collapses down to a black hole. But this doesn't mean it's gone; the squashed-down star is still there, and it's still pulling on the other star with its gravity. So the bright star still keeps going round it in orbit as before. But now when we look, we don't see two stars going round each other; instead we see a star going round . . . well . . . what? Nothing? That can't be right; that's not what stars do. We know it must be going round something – even if we cannot see what it is, that is to say, even if it's *invisible*. Not only that, but if we measure how fast the bright star is travelling round its orbit, we can work out how hard gravity must be pulling on it to keep it in that orbit. And that means we can work out how massive the invisible star must be. It turns out to be *very* massive – as heavy as you would expect a star to be if it had collapsed down to a black hole.

(I hope you're getting convinced! If not, read on . . .)

I have said how anything coming too close to a black hole gets sucked into it. That goes for the atmosphere of the star orbiting the black hole. Its top layers sometimes get ripped off and sucked into the black hole itself – like bath water going down the plug hole. And just as the bath water gets faster and faster the closer it gets to the plug hole, so the clouds of gas captured from the star speed up on the last stages of their journey. They get so fast and hot we know that they must have been speeded up by an absolutely enormous gravity – the kind of gravity you could only get with a black hole.

As I said, we get to know that there are black holes the same way as we learn that there is an invisible man. (Except, of course, there *isn't* an invisible man – he was in a made-up story. We don't *think* black holes are made-up!)

I am glad you are going to be a scientist, James. It can be exciting, as you say. It is also a lot of hard work – but that's how it is with anything worth doing, don't you think?

Time

66 When did time start?
McKenna Mills (aged 9)

The Universe was created in a Big Bang. By that we
do not simply mean all the stuff that's in the
Universe. It also marked the creation of space.
Recall how I was telling Farah (question 49) that we
scientists have some very peculiar ideas when it
comes to space. For most people, space (meaning
empty space) is just 'nothing'. But that's not the
way we see it. For us it is a 'something' — a very,
very smooth, evened-out 'something'. And like all
'somethings', it needed to be created. It started out
all squashed up as a dot with no size at all, and then
with the Big Bang, it suddenly expanded. It is still
expanding. That is why the distant galaxies are
rushing apart from each other; the space between
them is expanding. It is not that the galaxies are
moving through space; rather it is the moving space
itself that is carrying the galaxies along with it.

If you find that a difficult idea, try imagining a
rubber balloon with 5p coins glued to its surface.
Now blow up the balloon. What happens to the

coins? They move apart from each other. Is this because the coins are moving *through* the rubber, or sliding *over* the rubber? No. They move like this because the rubber in between them is expanding. That's how it is with space. The space of the Universe is behaving like that rubber sheet, and the coins behave like the galaxies.

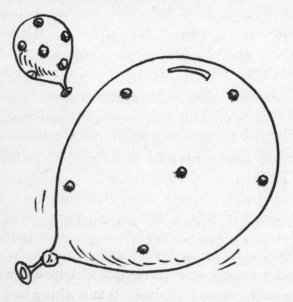

Now, what's this got to do with your question about the start of time? Well, just as the Big Bang marked the creation of matter and of space, it also marked the creation of time. In other words, the Big Bang marked the start of time! There was no time before the Big Bang. That is what today's scientists say. But they were not the first to come up with the amazing idea that there was no time before the world had been created. Someone got there before them; someone who lived 1,500 years ago and didn't know anything about modern science and the Big Bang. His name was Saint Augustine. The way he argued was like this:

How do we know that there is such a thing as 'time'? It is because things *change*. One moment the runners are at the starting line on the race track; at another moment these same runners are half-way down the track. What is the difference between the two? Time. What we are talking about happened at two different times. That's what we say. We can then invent clocks and watches which change in a regular way so we can measure the changes in time; we can say the difference in time is five seconds, or six seconds, or whatever it might be. That way we all think we know what we are on about when we speak about 'time'.

But, said Augustine, suppose nothing changes. Suppose nothing had *ever* changed. In such a world, would we know what time was? His answer was *no*. We would not even *be aware* that nothing was changing, because there would be nothing going on in our brains. (So, it's all right, you wouldn't be bored in such a world; you need time to be bored.)

So, a world where nothing changed would be a world where there was no time. And as for a world that had not even been created yet – so there weren't any things *at all*, let alone things that *changed* – obviously there could be no time. That was how Saint Augustine argued that there could have been no time until the world was created. Clever, don't you think?

Augustine is one of my heroes. Had he lived today I reckon he would not only have been a great saint, he would have won a pile of science Nobel Prizes as well.

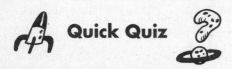

Quick Quiz

Can anything have existed before the Big Bang?

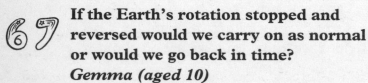

If the Earth's rotation stopped and reversed would we carry on as normal or would we go back in time?
Gemma (aged 10)

No, we would not go back in time if the Earth's rotation reversed. In fact, I don't think it will ever be possible to go back in time (despite what we are told in science fiction stories and films). Playing around with these ideas can be lots of fun, but if you take them seriously you soon get into problems. Suppose, for example, you suddenly got beamed back to an earlier time and found yourself driving a stage-coach. While worrying about how you are going to get back to the present, your mind wanders and you accidentally run over and kill your great-grandmother. If that happened she would not have been able to give birth to your grandmother, who wouldn't have given birth to your mother, who wouldn't have had you. So, you couldn't have gone back in time in the first place, because you couldn't have existed in the first (or is it the last?) place.

Mind you, if the Earth's rotation suddenly reversed, you would still not 'carry on as normal'. Just think: The distance round the Earth's equator is about 40,000 kilometres. The Earth rotates once every twenty-four hours. So that means someone standing on the equator is actually whizzing round through space at about 1,700 kilometres per hour (though of course it doesn't seem like it because everything at the equator is going at the same speed). Now, suppose the Earth suddenly stops. What's going to happen? All the loose things on the surface (people for instance) will carry on at 1,700 kilometres per hour, whereas hills and mountains won't. So they'd better watch out! And then there is all that loose stuff called seas and oceans. All that water is suddenly going to come ashore at 1,700 kilometres per hour.

I reckon it's a good thing that the Earth is likely to carry on the way it is for a long, long time!

I am writing to you because I have a problem. The problem is this: I know that huge telescopes look into the past. But if you had a huge mirror out in space and the mirror reflected the Earth, would a telescope on Earth see the Earth's past reflected in the mirror? Or would the telescope see the future through the mirror which could be the present? This is very confusing, don't you think? I am aged ten and am finding this rather difficult.
Paul Tench

We saw in question 46 that light takes time to travel from one place to another. So, whenever you receive light, it is 'old' light. In other words, it is telling you about the way things used to be at the place where it was given out. Because the speed of light is so enormous (light could travel five times round the Earth in the time it takes to say 'rice pudding'), in normal everyday life these time delays are so absolutely tiny we don't notice them. We assume that what we are seeing now tells us how things actually are now.

But in space, things are different because the distances are so much longer. For example, it takes light just over four years to reach us here on Earth from the nearest star. So when we 'see' that star at this moment in time, what we are actually seeing is how the star was four years ago. And the further off the astronomical object, the 'older' the light we receive. The light from a distant star might have set out so long ago that the star itself could have blown up in a supernova explosion since then. Here we are observing it in our telescope tonight, and it does not even exist any longer. As for light from the farthest depths of space, that has taken almost the whole lifetime of the Universe to get to us – some 12,000 million years.

Now to turn to your particular problem: you have to think of light leaving the Earth – let us say in the year 1900 – travelling to your distant mirror, and reflecting back to Earth so as to arrive at your telescope today. What that light will tell us today is what was happening on Earth in 1900 when the light was given out. So it gives us a chance to look into the past. It certainly cannot tell us about the future; there's no chance that by setting up a mirror in space you are going to be able to learn about next week's winning lottery numbers in advance, if that's what you were thinking! Bad luck!

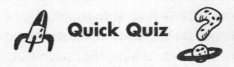 **Quick Quiz**

The Sun is 150 million kilometres away. If light travels at a speed of 300,000 kilometres per second, how old is sunlight when it arrives on Earth?

Is it true that there are dimensions exactly like ours only backwards? If so, is it like a rewinding video or looking in a mirror?
Katy Sigrist (aged 10¾)

Have you ever been to a film where you are looking at a person on the screen in close-up, then later, the shot widens to show that it was just a reflection in a mirror? It is only in the wide shot – where you can see both the reflection and the real person (facing the opposite way) – that you realize that you were fooled at first into thinking the close-up was for real.

Film-makers can use this trick because mirror reflections of everyday life look just as normal as the real thing. Of course, the writing on this page would look odd if you held it up to a mirror. But that is only because people decided long ago to form the alphabetical letters the way they do; they could just as easily have decided to write them all the other way round – back to front. (In fact, a famous scientist, Leonardo da Vinci, wrote all his notes backwards; his poor readers had to hold them up to a mirror to make them look normal!)

So, Katy, what this means is that if one lived in a part of the Universe where the space dimensions were reversed (left was right and right was left), it would not really affect anything; life would carry on as normal.

Much more interesting is your suggestion of the

time dimension being reversed. Here we are not thinking of ourselves as suddenly jumping back to an earlier point in time, and then living life normally (as I was discussing with Gemma just now). No, it would be a world where clocks run backwards; cars would start their journeys with no petrol, suck in exhaust fumes instead of give them out, and end their journeys with full tanks; and all of us would start our lives as an old person being dug up out of a grave, gradually getting younger, and finally ending up as a tiny baby being popped into mother. As you rightly say, it would be a world that behaved like a video rewinding.

This is not as stupid an idea as you might think. Scientists have seriously wondered whether the day will come when time (throughout the whole Universe – not just in one part of it) will go into reverse. That doesn't mean that you and I will repeat our lives – backwards the second time. No, it means that new and different people will, from that point on, live in that strange kind of world.

Now, for goodness sake don't go around telling your friends 'Uncle Albert says time will go backwards one day.' It almost certainly *won't*. It was just an idea, and nowadays most scientists think it is a rubbish idea. Besides, even if time did go into reverse, would we know it? No. The trouble is that in a world where *everything* goes backwards, your brain would work backwards, so your thinking would be backwards. If you could view a backwards-winding video with a backwards-thinking brain, the two 'backwards' would cancel out; everything would look normal! So, in a sense, time might already be going backwards and we don't know it! All a bit mind-boggling, don't you think?

Forces and Motion

How do aeroplanes fly in the sky?
Adil (aged 7)

First of all an aeroplane needs an engine. It sucks in air at the front, and throws it out at the back. In this way, the aeroplane is able to grab hold of the air, and 'pull' itself along.

So far, so good. But all that does is get us rushing down the runway; we're not off the ground yet. How does that happen? That's where the wings come in. The air pushes up on the underside of the wing, and that's what lifts the aeroplane into the sky.

But why does the air do that? Why does it push upwards and not downwards (or no 'wards' at all)? The answer lies in the cunning shape of the wing. If you look at the aeroplane sideways on, you might be able to see that the wing is curved: it's rounded on top, and flatter underneath. In fact, the pilot is able to change the shape of the wing by moving some loose bits at the back called 'flaps'. When these are pushed out they hang down and so make the shape of the wing hollow underneath. The effect of this is that as the aeroplane moves forward, the air above the wing fairly shoots past the rounded top surface, while that underneath tends to get caught up in the hollow. That way the air underneath the wing builds up and gets denser than that above. So, with more air underneath pushing up on the wing than there is air on top pushing down, the aeroplane rises up into the sky.

That's what they tell me – the aeroplane makers. And I'm sure they're right. It's a very sensible explanation. But I don't know. Every time I see one of those giant metal boxes lumbering down the runway, filled with people, baggage, duty-free booze, plastic meals, toilets, etc., I can't help thinking; 'This is stupid. It'll *never* make it. How can *air* hold up all that lot?' But it does! Weird!

71 Why can't an aeroplane go in space?
Daniel (aged 7)

The higher up you go, the thinner the air becomes. In the end you run out of air altogether. You are then in space.

But an aeroplane needs air to pull itself along. No air, no aeroplane.

That's why if you want to go into space you have to find some other way of doing it – a rocket. A rocket is a bit like a flying gun. With a gun, as the bullet comes shooting out of the end, the gun itself jerks in the opposite direction. The gun pushes the bullet forward, but the bullet also pushes the gun backwards. We say the gun 'recoils'. Of course, the gun doesn't come flying back into your hand at the same speed as the bullet goes forward (just as well!). That's because it's heavier than the bullet.

The same sort of thing happens in a rocket. What happens is that the rocket engine heats up some gas, and then throws it out. It pushes on the gas, and the gas pushes back on the rocket. The rocket 'recoils' in the opposite direction to that of the gas. So, if the gas goes shooting off to the rear; the rocket moves off in the forward direction. The gases are much lighter than the rocket (just as the bullet was lighter than the gun), so the gases have to come out very fast (they have to be hot) in order to push back hard enough on the heavy rocket to get it moving.

In this way, a rocket doesn't need to be in air. It carries its own supply of gases to push on.

 Why does the water in the bath go round and round when you pull the plug out? I want to know this because every Wednesday and Sunday I am puzzled by this, and my family too.
Rebecca Bullen

It is the same reason why an ice-skater spins faster as she pulls her arms in closer to her body. She starts by turning slowly with her arms stretched out, and then speeds up. This is because of her angular momentum. Angular momentum depends on how fast she spins (the number of turns per minute) and how spread out her body is. If she squashes herself down, then to keep the same angular momentum, she has to spin faster.

Now, what has this to do with your bath water? Well, when you get out of the bath, you set the water swirling about. As far as the plug hole is concerned the water is slowly rotating round it, either clockwise or anticlockwise. But at this stage the water is spread out over the whole bath. Eventually it all has to go out through the same narrow hole; it has to squash down, and that means . . . You've got it; the water has to spin faster to keep the same angular momentum.

But what if you got out of the water very slowly and very carefully? You would still find the water spins a little as it goes out. This is because, even without *you* swirling the water, it rotates very, very slowly because the Earth is spinning. The water spins one way in the Northern hemisphere, and in the opposite way in the Southern. Mind you, this is such a tiny effect, it can only be detected in a proper scientific experiment, where the water has been left to settle for *days* before the plug is pulled.

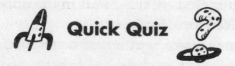 **Quick Quiz**

In question 40 I was telling Hannah about neutron stars. These are the squashed down remains of old stars. I said that they spun like crazy, perhaps completing hundreds of turns every second. Can you now see why they spin so fast?

73   What creates wind?
Matthew Clark (aged 10)

Wind is when the air around us pushes past us as it moves from one place to another. So the question is: why does air move around from one place on the surface of the Earth to another?

Near the equator where the Sun beats down from directly overhead in the middle of the day, the ground gets very hot. This in turn heats up the air near the ground. (Just look at the TV weather forecast and see how hot it gets in countries closer to the equator than ourselves.) In question 8 we saw how hot air rises. So that's what happens in sunny countries; the air travels upwards. But that's not to say it leaves a hole behind where it used to be. The surrounding cold air belonging to cooler countries like ours moves in from the side to take its place. That way we get winds at ground level. In time this air will get hot, it too will rise, and yet more cold air has to move in.

Once the hot air rises, it spreads out away from the equator as very high winds in the upper atmosphere. These, in turn, cool and come down in the colder parts of the Earth. So that way we get winds moving in a north-south direction between the hot equatorial areas and the colder areas on either side of the equator.

But things are more complicated than that. The Earth itself is spinning towards the east. (That's why the Sun rises in the east). This introduces a

123

spinning movement in the winds too, and that's how we get the great swirling movements of air you see on the weather charts.

Not only that, you can get local temperature differences. For example, when you are on a hot beach, you often get a wind coming off the sea. That's because the cooler air above the cold sea comes ashore to take the place of the rising hot air above the beach.

Add all these effects together, and what do you get? A mess. (Who'd be a weather forecaster?!)

How do clouds stay up in the sky?
Joseph Coleman (aged 8½)

Clouds are made of droplets of water. Water is heavier than air. So, you're quite right: the clouds ought to drop out of the sky! And yet they don't.

I have a confession to make, Joseph. When I got your letter, I was shocked. I didn't know the answer. A *professor of physics* and I hadn't a clue why the clouds stayed up there! I felt very silly.

But it wasn't long before I began to feel a bit better. You see, I walked down the corridor at the university where I work, and I asked nearly all the other physicists your question. (I pretended I knew the answer, and was just testing them.) And do you know what? Not one of them had a clue either! Oh, they came up with all sorts of ideas as to what *might* be going on, but it turned out none of them was right. Not only did none of us have the answer, it had never occurred to any of us even to ask the question. That's often how it is in science. There can be some problem sitting right under everyone's nose (or in this case, sitting above their head), and no one even notices that it is a problem. Then along comes some genius – like my hero Einstein – who becomes the first person to ask, 'Hey, what's going on here?' And then comes some big scientific discovery. Usually these really big discoveries are made by quite young scientists – those whose thinking is still lively and flexible, unlike we older scientists whose thinking tends to get stuck in a groove. That's why *you* came up with your question, and *we* didn't. (But don't get too excited; *someone* has already come up with your question, so you will have to think up another one in order to get your Nobel Prize.)

Now, just in case you're thinking I'm waffling on like this because I still don't know the answer, let me say I have now read a book on cloud physics and I now think I know what's happening.

Hot air rises carrying water molecules with it. The air cools and the droplets form to make a cloud. Because droplets are heavier than air, they start to fall through the air – which is what we expect to happen. But (and this is the important bit) the air they fall through is itself *still rising*. So in fact the droplets tend to be carried *upwards* with the upflowing air (though they don't go up quite as fast as the air, because they are falling through the air).

Right. Now you're thinking, 'OK, that explains why the cloud doesn't fall. But if the droplets are going up, why don't we see the cloud going *up*?'

The reason is that as fast as the droplets are swept upwards, more air and water molecules rise to fill the space they have left. It is now the turn of these water molecules to cool down and form water droplets in the same place as the first lot – before they themselves are swept upwards too. So, that way the bottom of the cloud stays where it is and seems not to be moving. But actually the cloud is continually *replacing itself*. As fast as old cloud moves upwards, new cloud takes its place.

It's a bit like what happens on the motorway. An AA man looks at the TV monitors and reports severe congestion between Junctions 6 and 8. An hour later he reports that the situation has not changed. As far as he is concerned, the TV monitors are showing exactly the same sort of picture. But that doesn't, of course, mean that he is looking at the *same* set of cars. The traffic *is* slowly moving; the cars he saw earlier have been replaced by another lot, but the shape and density of the traffic looks much the same.

Going back to those clouds; the upwards movement of the water droplets cannot go on for ever, obviously. What goes up, must come down – somewhere, some time. The upward moving cloud spreads out to the sides and eventually starts to fall. The air and its droplets are coming down. 'Ah! The cloud is at last about to fall out of the sky.' But no. As the air gets closer to the Earth, it warms up. And as it warms up . . . Yes, you've guessed it, the water droplets evaporate; the molecules escape the pull of the others and they drift off once more as invisible single molecules. So the bottom layer of that part of the cloud also stays at the same height. Whereas before the base of the cloud was where new cloud was always being made before it rose upwards, in this other part of the cloud, the base is where old cloud is being destroyed as it falls.

The next time you see a speeded up film of clouds moving across the sky, watch out for this. If you are lucky with some of these films, you will be able to spot this sort of thing happening.

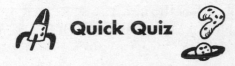

Quick Quiz

If the clouds are very thick, so the water molecules form really big drops, the rising air is not able to sweep them upwards. They are too heavy and they actually do fall. Also, even though these big drops are evaporating as they warm up on their way down, there can still be some of the drop left when it reaches the Earth's surface. What am I talking about?

75 **Why, when you are in the car and you are at the traffic lights, and then the light goes green, you go zooming off in the car and everyone goes flying back and bumps into the back of the seat?**
Eleanor (aged 9½)

You have to get used to the idea that motion is relative. While the car is getting up speed, you might think you are moving backwards – and indeed you are relative to the car. But that is not what you are doing relative to the road outside. Suppose some friends are standing on the pavement, what they see is the car suddenly going forward – and that includes the back of your seat. They see the seat going forward and pressing into you. As it does so, it sets you moving forward too. Once you are up to the same speed as the car, the seat back no longer has to push on you, and you are then able to sit normally.

So, are you going backwards into the car seat, or is the car seat coming forward into you? Which is it to be? To some extent it doesn't matter which. But, in fact, the second way of looking at it is better. After all, it is easy to understand how pressing your foot on the car's accelerator can make the car go forward; it's not so easy to see why doing that to the car should make you, the road, the houses, in fact the Earth and the entire Universe, go in the opposite direction!

I have been wondering for a long time what is absolute zero where nothing can survive.
James Webster (aged 11)

As you will know by now, the molecules of a hot gas jiggle about a lot. They rush around like mad, banging into each other and into the walls of the container holding it. As the temperature is lowered, things calm down and they move about less quickly. This in fact is what we mean by 'temperature'; it is a measure of how much energy the molecules have on average. The lower the temperature, the lower the energy. Normally we measure the temperature in terms of degrees Celsius (also known as degrees Centigrade). A hundred degrees Celsius (or 100°C) is the temperature of steam; 0°C is the temperature of melting ice.

But this is not to say you can't have temperatures less than 0°C. That value of zero was chosen simply for convenience. But how low a temperature can you have? Minus infinity °C? No. The point is that you can keep on lowering the temperature of a gas by taking more and more energy from its molecules. But eventually you get to a point where there is no more energy left in the molecules for them to give you. That happens at a temperature of about –273°C. So that is the lowest temperature you can have.

Knowing this, some scientists have decided it is more sensible to choose this lowest temperature as the zero of the scale. They call this 0 Kelvin (or 0 K);

this is what is meant by *absolute zero*. On this new temperature scale, ice melts at 273 K.

I have talked about a gas, but the same kind of thing holds for solids too. In a solid, the molecules don't have the freedom to move about from one place to another like those in the gas; they have to stay put, firmly attached to their neighbouring molecules. But they do have energy; they vibrate about their central positions. The higher the temperature, the more vigorously they vibrate. As with the gas molecules, by the time the temperature has dropped to absolute zero they have no more energy to give up.

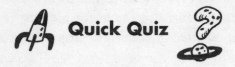 **Quick Quiz**

(a) What temperature would steam be on the Kelvin scale?
(b) Having learned what it is like at absolute zero, why can't life survive at that temperature?

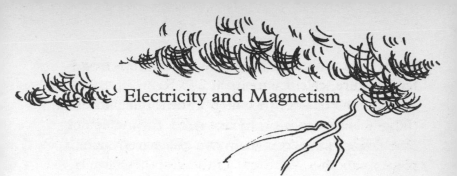

Electricity and Magnetism

How is electricity made? People can't just make it and it doesn't just appear from nowhere.
Caroline Atkinson

No, it doesn't come from nowhere. I explained to Lin (question 1) about atoms: how there was a nucleus at the centre of the atom and tiny electrons buzzing around the outside. We saw there that if an atom is on its own, the electrons are held quite tightly to their nucleus by the force between the negative electric charge on the electrons and the positive charge on the nucleus.

But when atoms are squashed close together (as they usually are) the electrons can get a little confused. They are still being pulled by their nucleus, but they are also being pulled by the nuclei belonging to neighbouring atoms. The outer electrons of certain kinds of atom can come loose; they are not sure which atom they belong to any more. They wander from one atom to another. This happens, for example, with the copper atoms of a copper wire.

And that's how we get electricity. When you

131

switch on an electric light, the loose electrons get pulled through the wires. Because each electron carries electric charge, there is a flow of electric charge. That is what we mean by an *electric current*. It's like the flow of water in a river. We call that a 'current' too.

It's hard work pushing the electrons round the wires. That's the job of power stations. They get the energy they need for this from burning coal, oil, or gas – or they might use nuclear power. Sometimes power stations can use water. If water collects in a lake high up in the mountains, it can be let out gradually and falls down pipes due to gravity. Its energy at the bottom of the mountain can then be used to push the electrons along.

 I would like to know what makes thunder and lightning and what makes it so loud?
PS Don't spend too long doing it.
Neil Butler (aged 7)

OK. I'll try and keep it short; I know how busy you seven-year-olds are.

Lightning happens when there are thick dark clouds about. Clouds are made of water droplets and tiny pieces of ice. As these move about, they rub against each other and some of the electrons get separated from their nuclei. The nuclei are swept upwards on the tiny ice crystals, while the knocked-off electrons hitch a ride on the downward falling

132

droplets. So there is a build-up of positive electric charge high up in the clouds, and of negative electric charge at the bottom of the clouds.

This cannot go on for ever. As you know, there is a pulling force between positive and negative electric charges. The electrons at the bottom of the cloud pull on the nuclei that have now been taken to the top of the cloud. Not only that, they pull on the nuclei of the atoms belonging to the ground. As the charges build up, these forces get stronger and stronger. In the end, something has to give. Quick as a flash (literally), the electrons and their separated nuclei suddenly rush back together again. Either that, or the electrons and the nuclei of the atoms of the ground make a dash to join up. And that is what we mean by lightning. It is the sudden surge of electric current in the air.

In this rush, the electrons bang into atoms along the way. Some of their energy gets changed into the energy of light. And that's what gives the jagged white line you see in the sky. It marks the path taken by the electric current. You can see it clearly as a line when the lightning strikes the ground – what we call 'forked lightning'. If the lightning is between the bottom and the top of the clouds it is likely to be hidden from us by the cloud. In that case, the lightning lights up the whole cloud, and we call it 'sheet lightning'.

Thunder is simply the sound that is made when there is this sudden surge of electricity through the air. It is loud because lightning strikes are very, very violent; they release a tremendous amount of energy. The reason we hear the sound *after* we see the lightning is that it takes longer for the sound waves to reach us than the light waves. Light travels at 300,000 kilometres per second, so reaches us in next to no time, whereas sound travels at a much slower speed of 332 metres per second (i.e. just over 1,000 kilometres per hour).

Lightning strikes more than a hundred times per second, every second, somewhere on Earth.

I think there is nothing more thrilling than watching a really good thunderstorm. (Though, to be honest they still scare me rigid – even when I watch them from the safety of indoors!)

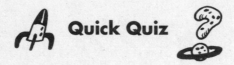 **Quick Quiz**

You see a flash of lightning and, using your watch, you time how long it takes before you hear the thunder. It arrives after twelve seconds. How far away do you reckon the lightning strike was?

79 I was thinking about what happens to a compass at the North Pole. Where does the arrow point? Since north on a compass is always facing the North Pole because of magnetism, would it stay on north? Also, how do people know where to go on expeditions on the North Pole?

Emily (aged 11)

When using a compass, you normally hold it so that its face points upwards. In other words, the magnet swings around in a horizontal plane. That way the needle points north.

But suppose now you have a special compass that is able to rotate in the vertical plane instead of the horizontal one. You might think that the magnet would lie parallel with the ground, pointing north-wards. In fact what you find is that the magnet points somewhat downward. And the further north you go, the more steeply downwards the magnet points. At the North Pole itself, it wants to point directly down. (The arrows in the drawing show the directions in which the magnet wants to point at different places around the world.)

This means, if you are standing exactly at the magnetic North Pole, and you hold the compass face horizontal (as is normal), the poor needle doesn't know what to do. It wants to point directly downwards, and you won't let it! So it just swings around aimlessly in any direction.

As for how North Pole explorers manage, I am not sure. I imagine they get their bearings from looking at the stars.

By the way, notice I talked of the *magnetic* North Pole. This is because it does not coincide exactly with the *geographic* North Pole – the one on which the lines of latitude and longitude (shown on maps) are based. The geographic pole remains fixed. But the magnetic pole tends very slowly to wander about a bit.

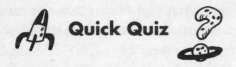 **Quick Quiz**

What do you think happens to the compass magnet as you approach the South Pole?

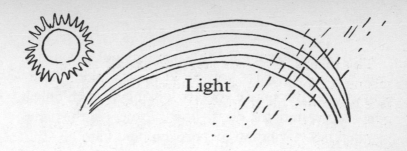

Light

I was flying off to Kos (a Greek island) and I looked outside and the blue sea was so dark! Then I thought, Why is the sea blue? Because when you pick up a handful of water, it's clear!
Gemma Harvey (aged 9)

One of the delightful things I remember about a holiday I had in Greece was the way the sky was always blue. And that immediately gives part of the answer to your question. The surface of the sea reflects the light from the sky, and if that is blue, then the sea will appear blue. If it's cloudy and grey, like it so often is in our country, then the sea looks grey and not nearly so inviting.

But the sea can also have a greeny-blue colour of its own. It depends on what salts and other chemicals have dissolved in it, and whether there is green algae floating in it. Mind you, this colour is only noticeable if you are looking through a lot of the water – if the water is deep and clear. If you have only a handful of it, it will still be a very, very pale shade of greeny-blue, but you won't notice it.

It is the same with glass. When you look through a window of clear glass, you could swear it didn't have any colour. But notice you are looking through only a shallow depth of the glass – the thickness of the pane. Sometimes, a window gets broken (and I am NOT suggesting you do this experiment!). It is then possible to pick up a large piece of the broken pane – very gingerly and carefully, of course – and look through it edgewise on (so you are now looking through a big depth of glass). You should now more easily see the true colour of the glass, which is normally green.

 There are many beautiful things in the world and I've always wondered about rainbows. How do they make rainbows? How do they get their colours?
Alexandra Cooper

As you get older you tend to take things for granted. But not rainbows. I still get excited when I see one.

To understand how they form I must begin by telling you about light – ordinary white light, like you get from the Sun. 'How boring,' I hear you say. 'I want to know about the *colours* of the rainbow, not about light that hasn't *any* colour.'

WARNING: Don't be fooled by your mum's washing machine with its labels saying 'coloureds' and 'whites'. Fair enough, it's important not to mix up white clothes with coloured ones when washing them. (I once forgot this and for months had to wear vests and pants that had gone pink!) But the amazing thing is that white light is actually the most 'coloured' light of all. It's a mixture of all the colours of the rainbow. How come?

Light is made up of waves. It has a series of humps and dips, just like waves on the sea. The distance between the humps is called the *wavelength* of that light. It's the wavelength that decides the colour of the light. For example, red light has a wavelength about twice as long as blue light, and yellow light is somewhere in between.

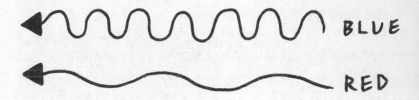

BLUE

RED

White light is different – it is a mixture of waves, all with different wavelengths. We know that because we can separate them out. How do we do that?

First we note that light travels at a certain fixed speed through empty space. It doesn't matter what the wavelength is; the speed is always the same. But this is not true when light passes through glass or water – the longer the wavelength, the faster the speed. That's good because we can use this to separate out the different wavelengths (or colours). We can pass the white light into a wedge-shaped piece of glass, called a *prism*. This slews the light round to one side so it comes out in a different direction.

How much does it slew round? That depends on how fast the light is moving through the glass. Fast

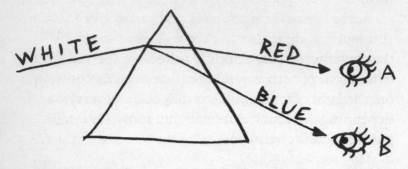

red light tends to charge ahead so it changes direction only a little; slow blue light slews round much more. Medium fast yellow light will be somewhere in between. You can see this happening in my picture. If you imagine putting your eye at point A, you will see the red light; at point B, the blue.

Now, the same kind of thing happens with water droplets. In the next drawing, you see white light from the Sun being split up as it passes through the rain droplets and comes back out again. I've shown you looking up at the sky. What do you see from that position? You are in line to receive the red light from the top droplet, and the blue from the bottom one. In other words, you see different colours depending on which direction you look. Different parts of the sky send you different colours – you are looking at a rainbow.

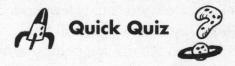 **Quick Quiz**

(a) Where in the picture of the prism would you have to place your eye to see the yellow light?
(b) Notice you need two things to be able to see a rainbow: the sun must be shining, and it needs to be raining. It's not often you get both of these at the same time. Why?
(c) I've talked only about the red and the blue light of the rainbow, but as you know, it has more colours than that. Can you name them?

 Why is the sky blue? If you can't answer it, that's OK.
Tracey Smith (aged 9½)
We have just seen one way of splitting white light up

into its colours. Here is another:

When sunlight passes through air it gets scattered to some extent by the molecules that make up the air, as well as by molecules of water and by dust particles. The direction it bounces off at depends on the wavelength of the light: the smaller the wavelength, the bigger the bounce.

When you look at the sky right above you, what you see is sunlight that has scattered through a big angle. So, it is mostly the short wavelength light you see – and that of course, means blue light. That's why the sky is blue.

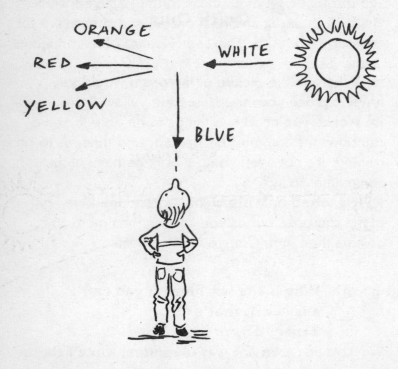

But, meanwhile, what has happened to the other colours: red, orange, yellow? This is where I can immediately go on and answer a question that you *didn't* ask, but one I'm sure you also want to know the answer to: 'Why are sunsets red?'

Like the blue light, the red light is also scattered by the air molecules – but not so much; it's more likely to go straight on, as you see in my picture. That's why you don't see it when you look overhead.

But now suppose the Sun is about to set. It is low down on the horizon, and that means its light has to go through a lot of air close to the Earth's surface, with all its smog and dust particles. So there is a lot of scattering of the short wavelength blue and violet light. That means only the reds, oranges, and yellows have much chance of getting through to you. And that's why sunsets look red.

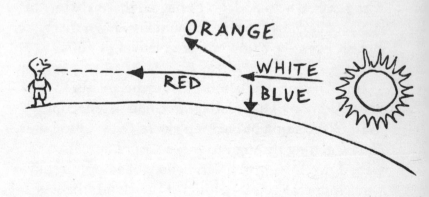

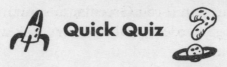

Quick Quiz

Why does the sky look black to an astronaut, instead of blue?

I was sitting looking at the sky when it suddenly turned grey, and started to rain. I ran in the house and wondered why does the sky turn grey when it is going to rain?
Hannah Ferguson (aged 9)

By an odd coincidence I happen to be answering your letter while flying in an aeroplane. I'm lucky today and have a window seat. Not only that, but it is not over the wing so I have a good view down below. That's the good news. The bad news is that it's cloudy. The pilot has just announced that we are flying over Greenland, so I imagine there are lots of spectacular snow-covered mountains down there. But all I can see below me is a blanket of fluffy cloud.

It is dazzling white. And that is the answer to your question. That blanket of cloud is reflecting most of the sunlight back up again. Only a little of it can be getting through to those mountains. If you were down there (brrr!) and you looked up, the sky would appear grey. The thicker the cloud, the less light gets through, and the darker the sky would be.

But of course the thicker the cloud, the more

likely it is to rain (or in Greenland, snow). So that's why it is a good idea to take your anorak when the clouds look dark.

 Why do 3-D things pop out at you when you're wearing 3-D glasses, but don't pop out when you are not wearing glasses? I want to be a scientist when I grow up so I can learn and explore different things.
Samantha Lindsay

As you look at this book you can tell it is 3-D; it looks solid. Now close one eye. The 3-D effect has gone. The book and everything else in the room looks flat. Next, open that eye and at the same time close the other. Keep repeating this: close left and open right; close right and open left . . . What do you see?

The book appears to jump about. That's because each eye is looking at it from a slightly different angle. And that's what your brain has to work on. It has to make sense of two slightly different, flat pictures. (Incidentally, you can open both eyes again!) In some absolutely marvellous and totally mysterious way the brain takes these pictures and produces a solid-looking 3-D book.

That's what happens normally. Now, let's *fool* the brain. Instead of getting the eyes to look at the solid book, we show one of your eyes a flat picture or photo of it, and at the same time we show your

145

other eye a different photo taken from a slightly different angle. The messages going from the eyes to the brain are exactly the same as they were before. The brain can't tell the difference, so you end up seeing a solid-looking book again.

But we can't just hold up the two pictures in front of you; each eye would see *both* of them. That's where the 3-D glasses come in. One lens lets through only green light, the other only red light. So, if we hold up a picture where everything is simply different shades of green, it will be seen only by the eye wearing the first lens; the other eye will see nothing. In the same way a red coloured picture will be seen only by the second eye. In fact, we can print both the green and the red picture on the same sheet of paper and hold it up in front of both eyes. Without the glasses on, the doubled-up pictures look a mess. But with the glasses on, each eye sees only 'its' picture – and out pops the 3-D!

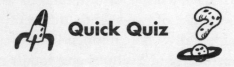 **Quick Quiz**

Hold your finger close up in front of your face. Look past it at a distant object out of the window. As you first close one eye, then the other, notice that your finger seems to change its position in relation to the background scene. Keep repeating the closing and opening of your eyes alternately while noting the extent to which the finger jumps from side to side. Now place your finger at arm's length, and repeat once more, still looking past it at the distant scene. Does the finger jump about more when it is close to your eyes or when it is further away?

 One day I was in my room listening to Criss Cross when all of a sudden a thought struck me. Why do we have sound?
Stewart Cromick (aged 10)

Suppose you are playing a CD, what happens is this: the middle part of the loudspeaker is made to vibrate back and forth. The higher the note being played, the faster the vibrations. The louder the note being played, the bigger the vibrations.

This vibrating loudspeaker pushes against the layer of air next to it, squashing it together. This squashed layer of air now pushes on the next layer, squashing that one. This then pushes on the next; that one pushes . . . etc. This carries on right across the room.

In the end, it's the layer of air next to the middle part of your ear that is being pushed. This part of your ear is called your 'ear drum'. In this way your ear drum is set vibrating. It is now vibrating in the same way as the loudspeaker was. When this happens, you hear a sound in your mind – the music being played by the CD. (Don't ask me why. No one knows why vibrating an ear drum should match up with a sound heard in your mind. It just does. Put it down to another of life's mysteries.)

And that goes for all sounds – not just those produced by loudspeakers. Whenever something makes a sound it disturbs the air, and this disturbance travels through the air and sets your ear drum vibrating.

You have to be careful not to listen to sounds that are too loud. They will set your ear drum vibrating so hard it might split, and that will damage your hearing. I have a son who plays in a pop group. About once a year he persuades me to go along to one of their gigs. I get worried when I notice that the sound is so loud my chest is vibrating in time with the music. What on earth must it be doing to my delicate ear drums? I now always take cotton wool with me. Much better, and safer that way!

TVs and Computers

I would like to know how pictures get on to TV screens.
Emily Grainger (aged 10)

If you look closely at your TV screen you will see the picture is made up of hundreds of horizontal lines. They are so close together, you don't normally notice them. They are made by a tiny spot of light passing back and forth. Wherever the spot shines, it leaves the screen glowing there for a short while afterwards; that's why we see lots of lines where the spot has just been, and not just the single spot. The spot starts at the top of the screen, and when it has finished doing all the lines and reached the end of the bottom line, it flicks back up to the top and starts all over again. As fast as the lines fade away, the spot returns to brighten them up again.

Now, if the spot were always the same brightness, we would simply get a plain white picture. But by changing the brightness of the spot, you can get it to 'paint' a picture. But how does the spot know how bright it ought to be at each point on the screen? That's where TV waves come in.

You have already learned (question 81) that light is made up of waves. The distance between the humps (the wavelength) tells us what the colour of the light is. We said how the wavelength of red light is about twice that of blue. But suppose the wavelength were even longer than that of red light. What colour would we get? The answer is: none. The 'light' becomes invisible! But although our eyes are not sensitive enough to see it, it is still there. We call it *infra red* radiation, or 'heat' radiation. When we feel the heat coming from a fire, it is the infra red radiation that is warming our skin. If the wavelength is stretched out even further we get other forms of radiation: microwave radiation (as used in microwave ovens for cooking), and waves used for sending radio signals, mobile phone messages and TV pictures.

The TV broadcasting station codes a message on to the waves it sends out. Your TV aerial picks up the waves and feeds them into your set. There the electronics decodes the message. It is the message that controls the brightness of the spot as it passes across the screen. And that's how the pictures appear.

I don't know about you, but I find it quite a thought that the room I'm sitting in at this very moment (like yours) is full of invisible coded messages flying about: all the different TV channels, all the radio broadcasts, and all the mobile telephone messages. We would never know if it wasn't for our TV set, our Walkman, or our phone.

How can computers remember every-thing?
Kathrine Jackson (aged 10)

When it comes to remembering things, computers are like libraries. They store the information away in much the same way as books are stored on shelves. Each book is coded and the shelves labelled. A good librarian knows exactly where to find each book when it is needed. And that is how it is with a computer. All the information is stored away. But more than that, there is a system for knowing where each bit of information was put, and how to find it when it is needed again.

No one fully understands how our own memories work, and why they are sometimes so bad (mine is awful and gets worse the older I get). One idea is that we have far, far more information stored away in our memory than we ever realize. It's all in there somewhere in our mental library; it is just that we don't have a very good system for finding it – we keep losing track of where we put it! That would explain those occasions when our memory is jogged by something we see or hear, and suddenly we remember some little thing that one would have thought had long ago been forgotten. But it hadn't been forgotten; it was there all the time.

Computers can get a bit scary. They have these amazing memories, and they can do sums and sort out information incredibly quickly. And, of course, that can be very useful for us. But apart from that, they are pretty dumb. After all, they're only doing what *we* tell them to do, so who really is the clever one? Mind you, whether it will always remain like that I don't know. No one can tell what the future holds. Perhaps in the end it will be the computers who will be telling *us* what to do. But we are still a long way from that. If the worst comes to the worst and they get really bossy, I suppose we could always pull their plugs out!

Life Processes

I am very interested in things like evolution. My question is: What is life?
Amy Firmin (aged 10)

People have wondered about this for a very long time. What is the difference between living things like plants, animals, and humans on the one hand, and non-living things like air, rocks, and rivers on the other? They are all made out of the same kinds of atoms, so does that mean living things have an extra ingredient, an extra something that non-living things don't?

No, that doesn't seem to be the answer. In fact, biologists can find it quite difficult at times to decide whether certain things, like viruses, are to be classed as living or not. There is no single deciding factor. Instead, it is a case of looking for a *combination* of factors. For example, living things tend to grow – at least at the beginning of their lives. They take in substances from the surroundings (eat food and breathe air) and use them so as to be able to grow and get energy for all the things they do. They reproduce (have babies). They get rid of unwanted substances (go to the toilet!). They can respond to their surroundings (find food, avoid dangerous enemies, know when something is too hot or too cold to touch, and so on).

Non-living objects might behave in one or two of these ways. A skin on custard will grow as it cools, an iron nail will respond to a damp atmosphere by going rusty, or a fire will keep burning by taking in fuel and air, before it eventually 'dies out'. But they don't show *all* those behaviours.

The fact that the dividing line is not all that clear is a bit messy. But it is important that it is like that. It meant that in the far off past we could go from a world that had only non-living chemicals on its surface (soon after the Earth formed) to the one we have today with all its different animals and plants. They developed through evolution. I am glad that you are interested in evolution, Amy. It is an absolutely fascinating subject, and I'll be saying more about it later (questions 93–99).

 I would like to know how come I am getting old?
Rachel Potter (aged 10)

Oh dear! Are you worrying about that already?!

At first, growing old is fun. It's something to look forward to. As each year goes by you get bigger and stronger. You can do things you couldn't do when you were little. You no longer have to tell fibs about your age to get into cinemas. Bullies at school now think twice before picking on you. As you go from being a child to an adult, you discover that you are in charge of a wonderful body – a machine that is far, far more marvellous than any machine we scientists can ever put together.

But then in your late teens and early twenties, things start to change. Your body, like any other body that has to keep going non-stop, starts to show the first signs of wear. Perhaps your eyesight begins to get worse; or you don't hear as well as you used to; later, your knees start to creak and it becomes uncomfortable to walk; or you notice that when you cut yourself, it takes longer to heal.

No one likes these sorts of changes. Luckily there is often something that can be done about them. For example, I now wear glasses to read, and I have a hearing aid. With these I am able to carry on much the same as I have always done.

Getting old is nothing to be frightened of. It happens to us all. The only way *not* to grow old is to die young – and who wants to do *that?!*

There is a saying: 'You are as old as you think you are.' That's very true. I sometimes think that, in my mind, I never really grew up completely. I feel a bit like a boy wearing a grown-up's body, pretending to be grown-up.

There's something special about being old; you have memories to look back on. You have learned so much from experience. You know more than you ever did before. You hardly ever have to look up the spelling of a word because you already know it – how about that! I used to look forward to being fifty. 'I will be at my best when I am fifty', I used to say. I was wrong. I am at my best *now* – when I am sixty-five – and it's getting better all the time. I can honestly say that I am happier now than I have ever been, and I wouldn't swap with anyone younger.

I went to see a cousin. His mum said that I had grown so much she almost couldn't recognize me. After she had said that I started to think how bones grow with our bodies. Could you tell me how bones grow?
Daniel Towers (aged 10)

I'm afraid I don't know very much about this. I never studied biology at school. But I showed your letter to a friend of mine who is a biology teacher. What she told me was that a bone, like the long bones in your legs, grows at two places. These are 'bone-making factories'. (She didn't actually use those words – she used big biology-type words I can never remember.) These 'factories' are close to each end of the bone.

Bone, like everything else in your body, is made up of round cells. You can think of cells as another kind of Lego-block. Cells sometimes split in two. Each half then settles down to being round and grows to the same size as the first cell. So now you have two cells instead of one. These then split to give four cells, the four become eight, and so on. That is what happens in the 'bone-making factories'. More and more cells are made, and this gives

Bone-Making Factories

more and more bony stuff. That way, the bone grows, you get taller, and your cousin's mum gets amazed when she sees you.

Why do children grow, and then when they become grown-ups, they stop growing? Also why can't we see people growing?
Helen Andersen (aged 10)
The reason why you eventually stop growing is all to do with the many changes going on in your body as you go from being a girl to being a woman. Your body makes juices called 'hormones'. And these go around the body at that time, making changes so as to get you ready for having babies. In particular, the hormones recognize that you are big enough by now to have babies, so the bone-making factories can be shut down; you don't need to be made any bigger. (In any case, who wants to be a giant?)

Mind you, the body is very clever. Although it shuts down these factories, it still has ways of making new bony stuff as and when it's needed. So, if you have an accident and break a leg, new bony stuff can be made to patch up the break.

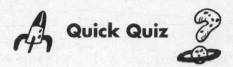 **Quick Quiz**

You asked me why you can't see people growing. Perhaps you could have a go at that question yourself if I give you a clue: you are in an especially boring school lesson and you keep looking at the clock on the wall. Do you see its minute and hour hands moving? Why not?

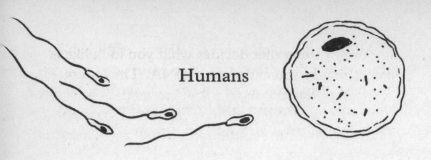

Humans

Hope you are well and your brain is ticking over as well as it usually does. Why do we look like our mums?
Jayne Braybrook

As you might already know, you were made out of a tiny bit of your dad, called a sperm, and a tiny bit of your mum called an egg. When these two joined up inside your mum, you were on your way! You grew and grew. Each day you looked less and less like a blobby egg, and more and more like a proper baby. After nine months tucked out of sight in your mum, you were ready to make your appearance: Birthday Number Nought. That was the signal for everyone to start saying things like: 'Isn't she adorable; she has her mother's eyes, her grandmother's mouth – her father's bad temper, etc.'

And it's true. You *are* like them. It's only to be expected. After all, you were made from a piece of your mum and a piece of your dad; and they in their turn were made out of pieces taken from your grandparents.

In fact we now know a lot more than that. The

important thing that decides what you look like is something in you called your DNA. The letters 'DNA' are the initials of a long name the biologists have given it. I can never remember these complicated names. But with this one it's all right because even the biologists just call it 'DNA'.

DNA is made up of a long chain of smaller molecules. (Remember a molecule is a collection of atoms stuck together.) Now the interesting thing about this chain is that it makes up a coded message. A secret code! Each of the small molecules is something like a 'letter', and they are grouped together to form something that we might think of as 'words'. The order in which these words appear along the chain then tells the body what it should look like: 'tall', 'blue eyes', 'brown hair', 'genius brain', etc. One of the most exciting things happening today in science is the way biologists are unravelling the code – finding out what each bit of it means. It's a big job. A human's DNA has the same number of 'letters' in it as you will find in 3,000 thick books.

Where did your DNA come from? It was made from copies taken from your parents' DNAs. That's why you end up looking like them with similar colour eyes, hair, etc. – you were built from the same set of plans. But your DNA is also different from everyone else's. It is a mixture of *two* people's DNA – your mum's and your dad's. That's why you are not an exact copy of either one of them. If your mum's DNA says 'green eyes' and your dad's says 'brown eyes', one of them is likely to win out – you will end up either with your mum's eyes or your dad's. But when it comes to the shape of your nose, or how tall you are, it might be the other parent's code that wins out. Not only that, but mistakes can happen as the original DNAs are copied, or your DNA might get changed as it meets up with chemicals in your body or is affected by radiation. Any of these will give rise to a brand-new code. That's why there is only one you; why we are all different.

Do you know how people were made?
Peter Jackson

Most scientists believe that you and I, and all the animals, were made through something called 'evolution'. Let me try and explain.

Let's begin by thinking of cheetahs. They can run very, very fast and that makes them good at chasing after zebras and antelopes, catching them and eating them. But not all cheetahs run at the same speed; some run faster than others. If there are not enough zebra to go round, which kind of cheetah will catch its dinner? The faster one. The slower cheetahs are the more likely to miss out and starve.

What makes some cheetahs faster than others? I was telling Jayne just now why children turn out to be like their parents. It's all down to their DNA. The same is true of the animals; they also have their own sort of DNA code – one that tells their bodies to grow into cheetahs rather than humans. Many things will make a cheetah strong or weak, fast or slow. One of these will be how much food it gets from its parents when it is small. Another is its DNA. Its DNA will play an important part in deciding whether it is likely to be a faster or slower runner than the average cheetah.

Those lucky enough to get the 'fast runner' code are more likely to catch their food. They will live to an age where they in their turn can have babies – babies that will inherit their parents' DNA, including the bit of code that says 'fast runner'. By accident, some will get a 'super runner' bit of DNA and outstrip everyone else. On the other hand, those cheetahs unlucky enough to have been born with the 'slow runner' code are in danger of starving to death. They are more likely to die before they have had a chance to have babies of their own. Because of this, the 'slow runner' code won't get passed on.

What this means is that there will be more members of the next generation with the 'fast runner' code than was the case for their parents' generation. We say that this code has been selected. The whole process is called *Evolution by Natural Selection*. No one is making the selection; it just happens naturally. The 'slow runner' code just dies out naturally, leaving the other.

In this way, we would expect the cheetah 'children' to be faster runners than their parents – on average. And of course, when it is the turn of these children to grow up and have children of their own, the same will happen all over again: those that happen by chance to have a DNA that helps them to run faster than the new average speed will be the ones that will survive and pass on that DNA to the next generation. In this way we can understand how cheetahs have got faster and faster with each generation.

The same will be true of anything else that might help an animal to survive longer and have a better chance of passing on its own DNA code: sharper claws, stronger beak, tougher protective shell, etc. That's how we come to have our modern-day animals. All the wonderful creatures we see around us today have gradually evolved over millions and millions of years from much simpler creatures. Each of them has developed its own special ways of surviving.

But, you might be thinking, what is so special about us humans? After all, we are not good runners compared to cheetahs, we are poor swimmers compared to fish, we can't fly like birds, we don't have sharp teeth or claws, we don't have a tough shell to protect us.

The special thing about the DNA of humans is that it gave us a big brain (for our body size) – a brain that can do very clever things. In the hard struggle to survive, our ancestors did not need sharp claws because they had the brains to design knives and axes; they did not need to run fast after deer because they could stand where they were and aim a rock or spear, or fire a bow and arrow.

It is not that we humans have a bigger brain than

any of the other animals; an elephant's brain is actually four times the size of ours. The point is that the bigger the animal, the bigger its brain needs to be just to keep the body ticking over properly. Because the elephant has such a large body, it needs a large brain just to keep going. The important thing is not how big the brain is, but how big it is *compared to the size of the animal's body*. It turns out we humans have a much bigger brain than you would expect for an animal our size. It's that extra brain size, left over after the body's needs have been looked after, that gives us the extra intelligence.

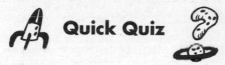

Quick Quiz

Many people find it difficult to believe in evolution. They say that it is impossible for all these developments to have happened purely by chance. Would you say that it was purely a matter of chance?

(i) Who invented talking?
Kathryn Ellison
(ii) Some countries talk different languages, right. Then how come people can't spell different? Who invented spellings? And why?
Danielle Dean (aged 11)
(That's right: two questions. As they say in the

supermarkets 'BUY ONE; GET ONE FREE'. Why not? I'm feeling generous!)

It takes brains to talk. In fact, there is a special part of our brain, on the left-hand side, that's in charge of talking. In the course of evolution, the brains of our early ancestors changed from being quite small to the size they are today. So the first clue as to when they started talking comes from measuring the size of their skulls. The other thing they needed before they could talk was a nose and throat area that was the right shape, and flexible enough to produce a rapid string of complicated sounds. Putting this all together, biologists reckon that talking might have begun about 100,000 years ago.

Of course, lots of animals make noises. Wild chimpanzees make thirty to forty different noises. Each has its own meaning, and says something important to the other chimps. I suppose you could call that a kind of 'talking'. With a lot of training, chimps can get across more complicated messages, and answer questions using sign language rather than voice. But it still doesn't add up to much compared with what we modern humans can do.

Like the chimps, we use just a few sounds. The English language is made up from just forty-nine different sounds (a bit like the sounds a young child makes when they start to say the letters of the alphabet: aah, ber, ker, der, eee, . . .) But, unlike the sounds chimps make, these don't themselves mean anything! The important thing is that we put these sounds together in different ways to make *words*. It's the words that can mean something. If I say 'table', then you have a pretty good idea what I'm talking about. But not all words have a simple meaning. If I just said to you the word 'but', or the word 'not', you would wonder what I was on about. But if I put them into a sentence, then they do make sense. In fact, it is the *sentences*, made up of strings of words, that really make the sense.

No one knows how our ancestors hit on this brilliant idea of replacing 'one sound = one meaning', by this way of stringing together meaningless sounds to make something that does have meaning. But no matter how it was done, it was probably the most important thing ever to happen to our ancestors. Talking makes all the difference between us humans and the other animals. It means we can learn from each other. When Mum says 'Don't touch that pan; it's hot!' you just take her word for it; that way you don't get burnt finding out.

And not only can you learn from people you meet today, you can also learn from people you have never met; indeed, from people who are now dead! How? Through writing. Writing is a kind of 'talking' using books – like the one you are now reading. And that is where spelling comes in. Whereas talking is all about stringing together a lot of separate sounds in a special order, writing is based on stringing together a lot of separate squiggles (letters of the alphabet). The order in which the squiggles are set down is a code that tells someone else what sounds you would have to make in order to speak the sentence. Good spelling is all about getting the code right so that people know exactly what you are on about. That's why it is important to learn to spell.

Mind you, having said that, it's a total botch-up job! For example, think of the different ways of pronouncing 'ough' ('thought', 'tough', 'the bough of a tree', etc.) No wonder you, and every other child in every other school in the country, complains. Any one of us could have thought of something more sensible. But that's the trouble. No one person did think up spelling. It just sort of grew, and evolved on its own.

Still, it does work. As I was saying, talking and writing is the way we learn from each other. Just think of all the things you know. Your mind is stuffed full of facts. But how many of those facts did you actually discover for yourself? Very, very few. Most of the information we have about the world is second-hand. It's what thousands and thousands of other people have discovered, and passed on to us – through talking and writing.

95 Why do humans rule the world?
Hilary (aged 11)

I get a bit worried, Hilary, when people describe us humans as 'ruling the world'. I think we have to be a bit careful when we say things like that. Certainly we are very powerful. We owe that to our brains, our ability to talk and learn from each other. That was the secret of our success as an evolved creature.

My worry is that being clever is not the same as being *sensible*. You might know someone at school, for example, who is top of the class and comes first in all their exams, but is as thick as two planks when it comes to how they live their life. That's how it might be with us humans. We have this wonderful intelligence, yet look what we are doing to the planet. Look at the way we fight each other in wars. With all that cleverness we have invented nuclear bombs, and now we have built enough of them to destroy everyone in the world in a matter of minutes. It's clever, but it's not sensible. Perhaps we are too clever for our own good.

Dinosaurs were around for 200 million years. We modern humans arrived a mere 100,000 years ago. Will we still be around in 200 million years' time? I doubt it. If we're the 'rulers of the world', I reckon the dinosaurs were better at it than we are; at least they ruled longer than we are likely to.

When it comes to being a successful animal there are far more insects than there are humans. And when it comes to long-surviving animals, you can't do better than bacteria. They have been around for 3,000 million years, and there are more different types of bacteria today than at any time in the past. Not only that, if there is to be a global nuclear war, it won't be the complicated animals like ourselves that will survive, it will be the tiny, simple, bacterial bugs. Perhaps we should think of bacteria as being the true rulers of the world; they ruled in the beginning and they will carry on until there is no more life on Earth.

Not that I would want to swap places with a bug! No, it's good being a human. With our big brains we are able to live much more interesting lives than any of the other animals. But we do need to be careful.

96 Why do people die?
Marvyn (aged 7)

Some people, of course, die as a result of accident or disease. But I imagine what you are wondering about is why people die from natural causes.

I suppose the main reason is that they just wear out. Take the heart, for example. It's a pump for pushing the blood round the body. It pumps about once a second. You can check that out by feeling the pulse in your wrist. (I have such difficulty finding my pulse, I sometimes panic, thinking I must have died already without knowing it.) That means by the time you get to being seventy years old, it must have pumped 2,000 million times. No wonder by then it's about to pack up. I reckon it does wonderfully well to keep going like that for all that time, without a break, and without a service or overhaul. I only wish man-made machines and electrical goods were as reliable as that.

When you think of how we humans got here (through evolution) I suppose it's a good thing we do die. If we didn't, then all those early ancestors of ours would still be around, taking up food and space, and not giving a chance to later generations to develop and become more interesting creatures – like ourselves. It's a strange thought that death is actually an important part of life. From the point of view of evolution, it is as necessary for all forms of life to die off, and get out of the way so as to make room for the younger generation, as it was for them to get born in the first place. In fact, it is now thought that just as there are codes in our DNA that govern how the different parts of the body should be built, there might also be codes that tell our body how and when it should die.

Animals

97 **If apes developed into humans, what developed into apes?**
Alex Marks (aged 10)

We think modern-day apes, like chimpanzees and gorillas, evolved from the same ape-like creatures that gave rise to us humans. You are asking what came before that. The answer is a small insect-eating animal. That in turn was developed from a reptile, and before that a fish, and before that little bugs like today's bacteria. We can't be absolutely sure, but it is very likely that it goes all the way back to slime and scum and chemicals in the sea! In other words, all living creatures (including plants and trees) have evolved from stuff that in the first place wasn't even alive. (Recall what I was telling Amy, in question 88, about there being no sharp dividing line between the living and the non-living.)

The whole thing probably began when some atoms came together in the sea to make a molecule that was able to make copies of itself – rather like DNA today makes copies of itself. That was the first important step. Once that happened, the first of these molecules became two, these two each copied themselves to become four, the four became eight, and so on. The other important thing was that there were mistakes in the copying. That way you got different varieties of the original molecule. (Normally we think of mistakes as being bad – we lose marks for them in an exam. But it's a good thing these 'mistakes' happened, or we would never have been here.)

Some of these different versions had a better chance of surviving to make copies of themselves than others. These were the ones that survived and developed further while the others died out. The surviving ones became bacteria. At some stage molecules collected together to form cells, and the cells together made up the bodies of animals and plants big enough to be seen down a microscope. Over time, these became ever more complicated, until in the end we humans and the other modern-day animals arrived on the scene.

All this must have taken a very long time. But we know that there has been a very long time in which this could have happened. The Earth formed 4,600 million years ago. The first bacteria appeared 3,000 million years ago. The first organism with more than one cell – so-called multi-celled organisms – appeared 1,000 million years ago. The brains of our ancestors began to grow about 2 million years ago, and modern humans arrived about 100,000 years ago.

98 Could you tell me why cats eat mice?
Jason Pickford

I have a cat named Curry. (No, no, that wasn't the reason. It was her colour, of course!) I love her dearly, especially when she rolls on her back and makes eyes at me. But there are times when I can't help flying into a great rage at her. That's when she goes out into the garden and kills birds and mice. She brings them into the house and I see red. After all, I know her tummy is full of Whiskas; she's not hungry, so why kill innocent creatures?

When I calm down again, I realize I'm being stupid. Curry can't help behaving like that. She doesn't *mean* to be cruel. The point is this:

You remember how I was telling Peter about the way we and the other animals evolved (question 93). When it comes to surviving, we were the lucky ones to get the big brains, or fast running legs, sharp claws, tough shell, and so on. These are all coded into our DNA. But DNA does more than provide a plan for how the body should be built. After all, it's no good a tiger having sharp claws if it doesn't know what to do with them. An animal that has an in-born tendency to *use* them – an animal that kills on sight – is more likely to get food when it is in short supply. Another without that tendency – one that has to think it all out from scratch every time – is likely to miss out on the meal.

So the DNA not only has codes in it for building the animal's body, it also has codes that tell the animal how to *behave*. And that is what is happening to Curry. On seeing a bird or a mouse she instinctively tries to kill it. She can't stop herself. It's her DNA that is telling her to do it. It was all part of her ancestors' survival kit. If her ancestors had not had this tendency built into them, they would not have survived, and Curry wouldn't be here today. The fact that those ancestors *did* have that code means that Curry has it as part of *her* code also – even though, since the invention of Whiskas and kind owners like me, there is no longer any *need* for it. These days it would be a better plan for cats to learn how to roll around and make eyes at humans.

Now, I don't want you to think that the kind of behaviour coded into the DNA is all about killing; it isn't. For instance, there is a code there that makes mothers feel especially protective to their children – even to the point where they will sometimes unthinkingly sacrifice themselves for their children. Then there is a code that makes the baby kangaroo, immediately it is born, head up mother's fur to find the warm, soft pouch that is to be its first home. These are all ways of helping the young to survive.

Lastly, one of the really fascinating questions is whether the behaviour of us *humans* is also influenced in this way. After all, we too are evolved animals. Of course, there is one big difference between us and the other animals. Because of our intelligence we can look ahead and plan ahead. We can work out *different* ways for us to behave. We can *choose* to go against our natural instinct, if we think that would be best. So, much more than the other animals, we go through life planning and making our own decisions. It can be far more complicated living a human life than that of any other animal – but much more interesting.

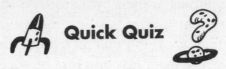

Quick Quiz

In what ways do you think we humans might show behaviour that is genetically influenced?

Why can't dogs marry cats?
Simerjit (aged 9)

The quick answer, I suppose, is that they don't fancy each other!

We say that they belong to different *species* (pronounced 'spee sees'). Animals that belong to the same species can generally mate and have children, but they can't with animals belonging to another species.

When we look back over evolution, what we find is that, over a long period of time, a particular species begins to split up. Some of its members develop one way, the others a different way. One sort might be good at surviving because they are gradually developing stronger and stronger legs and so can run away from enemies faster; the other might find it is developing a better grip and can climb trees, and get out of danger that way. After many generations, the two lots might be so different from each other, that they no longer like the look of each other, and even if they did mate, they find it less and less likely that they will have children. Eventually they find they can only have children with animals of their own kind. At that point we say the original species has become two.

In fact, we think that all living creatures originally came from just one sort of ancestor. That's because all DNA codes are similar. Obviously there

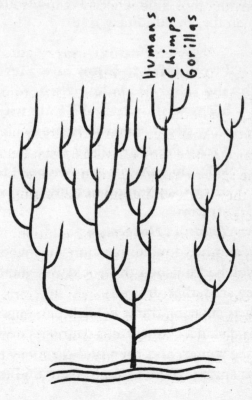

are differences in the DNAs of different species; if not, we would all look alike. But much of the code is exactly the same. That would be very unlikely to happen if we did not all share the same ancestor who originally had that particular coding. So, starting out from one species, this has branched out into different species, which themselves have branched into others. It's a bit like the way a tree starts out with just one trunk coming up out of the ground. That splits up into branches; the branches then split into thinner branches, which in their turn become lots of twigs. All the animals we see today are different twigs on the 'evolutionary tree'.

 I would like to know how pigeons and other birds know where to go because they travel a long way and they don't seem to be very clever.
Vicky Peplow (aged 10)
Why do salmon know where to go after they have been swept downstream?
Thomas Butchers

(Oh dear, another double question. Just when you thought you had finished the book! But you have to admit they are rather similar.)

It really is amazing the way birds can migrate enormous distances to warmer countries down south in the winter, and then find their way back up north next spring. If it was up to us to find the way we would need to carry a pocket compass with us. That's how we would keep track of north, south, east, and west.

In fact, that is exactly how the birds do it! They don't actually *carry* a compass with them, of course. They don't need to; they've got one already in their brain! There is a part of their brain that is affected by the Earth's magnetism – just like the little magnet you find in a pocket compass. We know this because experiments have been done where a small magnetic instrument was strapped to the bird. The bird picked up the instrument's magnetism, as well as the Earth's, and got completely lost!

So, magnetism is one way they do it. But they also have another trick up their sleeve (not that they have sleeves, but you know what I mean). It's a trick of the light. There is something very special about the light coming from the sky. To us humans it looks pretty much the same whichever direction it comes from. But not to a bird.

Light is made up of waves. Usually when it comes towards you, the tiny waves wobble, or vibrate, up and down and from side to side. But with light from the sky it is different. The sunlight scatters off the air on its way to us, and this can reduce some of the wobbles; it might now be vibrating up and down only, or side to side only. We say the light has been *polarized*.

Normally, we humans can't tell the difference between ordinary light and light that has been polarized. We have to wear polaroid sunglasses to do that. The polaroid lenses are arranged so that only the up and down vibrations get through. That means when you're looking at polarized light vibrating side to side only, it gets cut out; it can't get through the lens. (That's a good thing because sunlight reflecting off the surface of the road, say, is polarized side to side, so the glasses cut out that kind of glare.)

When you're wearing polaroids, if you look up at the sky, you will notice that the glasses cut out different amounts of the light, depending on which direction you are looking (even when it's cloudy). That shows the light is polarized.

The remarkable thing about birds is that they seem to be wearing polaroids all the time! There is something about the way their eyes are made that helps them to tell which way the light is vibrating – and that in turn gives them a clue as to which direction they are flying. So that's a second way birds can be sure of finding their way over long distances.

Of course, when they eventually get close to home, they no doubt look around and start to recognize the place anyway.

Now for *your* question, Thomas.

Having just read what I told Vicky about how birds find their way around, you might think salmon also have magnets in their brains. That might well be true. Scientists are not all that sure at the moment.

But one thing we do know: salmon are very good

at knowing what kind of water they're in. If you or I tasted some water taken from a river, lake, or sea, we could tell whether it was salty or not, and if it was salty, whether it was 'very' salty, or only 'slightly' salty. But salmon do much better than that. They can recognize different kinds of salt and other kinds of chemicals. And that's useful, because there are different amounts of different chemicals dissolved in the river or sea wherever you happen to be.

So, when salmon are trying to find their way home they test (or 'smell') the water they're swimming in. Once they start to recognize the chemicals, they know they must be getting close to home.

They might also be guided by changes in the temperature of the water. Water warmed by the sunlight as it flows along a river is likely to be warmer than the deep oceans into which it flows. So, a salmon out to sea looking for the entrance to its river, heads to where the water is warmer.

Not only that, it can obviously tell which way the water is flowing. So, it might recognize the pattern of ocean currents. Once it finds the entrance to its river, it knows the river is likely to be flowing towards the sea. This means, to get up river, it needs to swim in the opposite direction to the current.

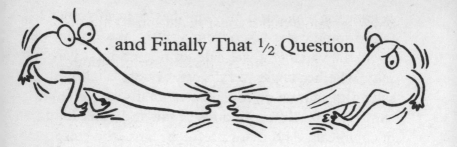
100 ½ Why does gravity exist?
Eleanor (aged 7)

I haven't a clue! Usually I don't like owning up to not knowing the answer. But this question is different. No one knows the answer – and that is how it is likely to be for all time.

The point is that scientists have the job of understanding and describing the world we live in. But that doesn't mean we can explain why the world is *this* one, instead of some other kind of world. So, I can tell you how strong gravity is, and how it changes with distance. In other words, I can tell you how gravity *behaves*. But I cannot tell you why we have gravity in the first place. Nor can I say what gravity *is*.

The same goes for time; I can measure it, but I cannot say what it is, or why it exists. The same goes for energy, for electric charge, and a whole host of other things. In a very deep sense, we scientists haven't a clue what we are talking about! And yet what we talk about does make sense – you have only to look at the way science has changed the world to know that it must make some kind of sense.

So, Eleanor, I cannot answer your question. I guess that makes the score:

Eleanor 10 Uncle Albert 0

But I don't feel badly about it, because it was not a scientific question. I know it *sounded* like a scientific question, but it wasn't one – not really. Science doesn't answer questions like that; that's why I counted it as only 'half' of a question.

Answers to Quick Quizzes

Some of the questions have a quick quiz. Here are the answers. Good luck with them. I hope you do well.

(HINT: When looking up an answer cover up the bottom of the page so that you don't accidentally (on purpose?!) catch sight of the answer to the next quiz before you have had a chance to have a go at it.)

Qu. 2

No. In question 1 we saw that it was the force between the negative charge on the electrons and the positive charge on the nucleus that kept the electrons in the atom. But the positive charge on the nucleus is due only to its protons; the neutrons don't have any charge.

Qu. 3

The word 'atom' means 'something that cannot be cut'. So, that would be a better description of a quark or an electron, rather than the atom to which they belong. But it's too late to change that now!

Instead, we call particles like quarks and electrons fundamental particles.

Qu. 5

We can't see them because individual molecules are so tiny. We can only 'see them' when there are a lot of them clumped together – as happens in a droplet or puddle. To understand why this is so, imagine looking at a distant field. You can tell it's made of grass. But at that distance you would not be able to see a blade of grass on its own. Only when all the greenery of lots and lots of blades are added together can you see and recognize what you are looking at.

Qu. 8

Not really. All she has to do is switch the kettle on to make a cup of tea! The water heats up, and the water molecules evaporate. As they come out of the spout they join the cooler air outside. Having started out piping hot, they now suddenly cool down and form droplets – a cloud of steam. Simple!

Qu. 11

It is true that two nuclei are not going to produce much energy. But when you get squillions upon squillions of them all doing the same thing at the same time, it adds up.

Qu. 14

Suppose we think of a patch of the paper the same size as the part of your eye that lets the light in (the black central disc called the pupil). It will receive as much light from the Sun as your eye would if the Sun's rays were allowed to fall on your eye directly. But what happens to this light? Some of it gets absorbed by the paper (that's why the paper warms up a bit) while the rest gets reflected. But note that it is not all reflected to your eye; it is scattered in all directions (you would not be able to see that patch from different angles if part of the light wasn't going off in those directions as well). From the particular angle you are viewing it, your eye collects only a tiny fraction of the total; that's why it is safe to look at the white paper. Mind you, the surface might still be uncomfortably bright, so it is better to wear sunglasses and cut down the glare still further.

Qu. 16

Even the centre of the Galaxy doesn't stand still; it is moving at an even greater speed of 600 kilometres per second through space. That is something else we have to know before we can work out the speed of the Earth.

But it's not just the speeds we need, we also have to know the directions in which they are going. After all, if a train moves forward at ten kilometres per hour (10 k.p.h.), and I walk along its corridor at six k.p.h., my real speed is 10 + 6 = 16 k.p.h. only if I

am walking towards the front of the train – in the
same direction as the train is going. If I am walking
towards the back, then my speed will be $10 - 6 = 4$
k.p.h. When we take note of the directions of the
Earth's motion relative to the Sun, the Sun's motion
relative to the centre of the Galaxy, and the centre of
the Galaxy's motion relative to the rest of the
Universe, we find that the Earth is speeding through
space at a speed of about 400 kilometres per second.

Qu. 17
Humans normally live about seventy Earth-years.
But Pluto's year is 248 Earth-years. So, if you lived
on Pluto, you would be dead long before your first
Pluto-birthday came round!

Qu. 19
How big the planets are depends on how much gas
was originally captured into each whirlpool, and
whether they were sufficiently far from the Sun to
be able to hang on to their gas once the Sun caught
fire and started blasting out its hot wind.

Qu. 22
If it took 2 million years to go from sixteen oversized nuclei to four oversized and twelve sensible-
sized nuclei, it will take another 1 million years for
half of the remaining four to decay, so leaving two
oversized ones and a total of fourteen sensible-sized
ones. So, a mixture of two oversized and fourteen

sensible-sized nuclei means it must have been around for a total of 3 million years.

Qu. 24
You will have to postpone your visit for 10 million years; that's how long it takes for the plates to travel the 600 kilometres distance between the two cities, at the rate of five centimetres per year.

Qu. 28
Every four hundredth year they have a leap year instead of an ordinary year (i.e. they have an extra day). That means the Earth must have got behind by a day.

Qu. 31
It is thought that the dinosaurs died out as a result of climate changes following one such massive impact some 65 million years ago.

Qu. 32
Because that's how long it takes the Moon to complete its orbit around the Earth and get back to where it started. As it goes round us, the light from the Sun strikes it at different angles.

Qu. 34
No. We get an eclipse of the Moon when the Earth casts its shadow on it. So the Moon, Earth and Sun have to be lined up as in my drawing. Only those

people who are on the side of the Earth facing the Moon are going to see the eclipse, and that means they are facing away from the Sun – so it must be night-time for them.

As for why you can't see an eclipse of the Sun at night, if you are able to see the Sun at all, it *must* be day time!

Qu. 37
No. There is little or no atmosphere out in space, so there is nothing out there to 'twinkle' the star-light.

Qu. 41
(a) They look bright to us because they are so close. They are only about a hundred kilometres away, whereas the nearest star is 40,000,000,000,000 kilo-metres away.
(b) No – because there aren't any in space. There is no atmosphere, so there is nothing to burn up the tiny dust particles.

Qu. 44

It takes three hours to travel sixty miles at twenty miles per hour. And that is the same time it takes to travel fifty-seven miles at nineteen miles per hour. So they must both have started out three hours ago.

In the same way, when we look at the speeds of the galaxies and how far they have travelled at those speeds, we can work out that they must all have started out together 12,000 million years ago.

Qu. 47

If the density of matter in the Universe has the critical value, then the expansion will stop, but only in the infinite future. So there will not be time for it to come back together as a Big Crunch. So, our best bet is that the Universe will end with a Heat Death.

Qu.52

All the 'loose' stuff on the Earth's surface, like water and air, would slide round to the other side, and fall off into space.

Qu. 54

It is sharply pointed at the front and smooth and sleek along the sides and bottom so as to cut through the water as easily as possible. This keeps water resistance as low as possible. That in turn means the engine does not have to work so hard to keep the ship moving and can save on fuel.

Qu. 57

The Moon is smaller than the Earth; its gravity is not as strong. (You have probably seen on TV how the Apollo astronauts were able to jump higher on the Moon than here on Earth.) The Moon's gravity was too weak to hold on to the jiggling molecules of air as they rushed about. That's how they escaped into space, leaving the Moon without an atmosphere.

Qu. 59

All its energy of motion gets converted into heat energy. As we have seen, the craft plunges through the atmosphere and heat is generated as the air rubs up against the craft's heat shield. So what started out as the energy of motion of the fast-moving spacecraft, ends up as the jiggling, random motion of heated up air molecules.

Qu. 61

When you walk, you put one leg in front of the other, pushing with your feet against the floor; that's what gets you going. But out in space there is no surface for the astronaut's feet to push against. So, no, astronauts do not really walk. Instead, they can grab hold of things and give themselves the odd push or pull to get themselves floating along. They can also carry little rocket packs with them. By squirting out gas in one direction, they recoil in the opposite direction.

Qu. 63

Normally we see an object either by the light it gives out (e.g. a star), or the light it reflects (e.g. the Moon). No light can escape out of a black hole, so the hole cannot give off any light of its own. Any light that falls on to it from outside gets sucked in and so does not get reflected. That is why no light ever comes from a black hole. And that's just another way of saying that it is 'black'.

Qu. 66

You can't use the word 'before' when you are thinking of the Big Bang. 'Before' means 'at an earlier time' – but there was no earlier time than the Big Bang. So, no. We cannot think of something existing 'before the Big Bang'. We cannot even think of 'nothing' existing before the Big Bang!

Qu. 68

The distance travelled is 150,000,000 kilometres, and this has to be divided by the speed of light, 300,000 kilometres per second. The answer is 500 seconds, or just over eight minutes.

Qu. 72

Like almost everything in space (the Earth, the Moon, and the planets), stars spin like a top – only very slowly. In the same way as a spinning iceskater has angular momentum, so do the stars. But if a star collapses down to a neutron star, only a few kilometres across, what is to become of its angular momentum? Well, just as a skater twirls faster when she draws her arms into her body, so the collapsing star twirls faster to make up for its smaller size. That's how neutron stars end up spinning so fast.

Qu. 74
Rain!

Qu. 76
(a) We have seen that between absolute zero and the temperature of melting ice there are 273 Centigrade degrees, and also 273 Kelvins. So a degree Centigrade is the same as a Kelvin. This means if the temperature of steam is 100°C higher than that of ice, then it is 100 K higher. So the temperature of steam on the Kelvin scale will be 273 K + 100 K = 373 K.
(b) At absolute zero, there is no energy left to do anything, so life is not able to carry on at such a temperature.

Qu. 78
The light arrives in next to no time, so we can take that to be the time at which the sound set out on its journey to us. The sound arrives twelve seconds later, so twelve seconds is the time it has taken for the thunder to get to us from where the lightning struck. Its speed is 332 metres per second, so the

distance must be 12 x 332 = 3984 metres, or about four kilometres.

Qu. 79

As you get closer to the South Pole, the magnet wants to point more and more upwards. At the South Pole itself, if the compass is held horizontally, then the needle swings around in any old direction; it cannot decide which way to point. On the other hand, if the magnet is free to swing in the vertical plane, then it will point directly upwards (whereas previously, at the North Pole, it pointed directly downwards).

Qu. 81

(a) Because yellow light has a wavelength in between red and blue, you would have to put your eye between the points A and B.

(b) To see a rainbow, we need clouds to give the rain. But it's clouds that tend to block out the sunshine. So we have to hope for a gap in the rain clouds for the Sun to shine through. If you have a hosepipe or garden sprinkler, of course, it's easy. You just choose a sunny day and make your own 'rain'.

(c) As for the colours of the rainbow, they are: red, orange, yellow, green, blue, indigo, and violet. I've never really been sure about 'indigo'. I reckon they just invented that one to give an initial letter 'i' so that the seven initials would make up a word that was easy to remember: ROYGBIV.

Except that even 'roygbiv' is not a proper word. Mind you, I have always remembered it, so it seems to work.

Qu. 82
Out in space there is no atmosphere to reflect the Sun's blue light to the astronaut, so the sky appears black. In fact, when flying in a jumbo jet at high altitude, if you look upwards out of the window at the sky above, you'll see it is noticeably darker than it was as seen from the ground. That is because there is less atmosphere up there.

Qu. 84
The finger jumps about more when it is closer to your eyes. So, when your brain compares the left and right eye pictures, it knows that the object that jumps about most is the one closest to you. And the more it jumps, the closer it must be.

This is how a stereoscopic camera works. It has two lenses at slightly different positions (instead of two eyes) and takes two slightly different photos of the scene. By measuring how much the different parts of the scene jump about in one picture compared to the other you can work out the 3-D shape. I had to use a computer when I was using one of these special cameras for my scientific work. The human brain is its own computer; it does it straightaway. Very clever, the human brain!

Qu. 91

The reason we can't see ourselves growing is that we grow so slowly. It's the same with the hands of a clock. If you watch the hands, they don't seem to be moving. It's only when you go off and do something else, and then come back to take a look, that you can see a change. That's how it is with children growing.

Qu. 93

The changes that happen to the DNA are indeed purely down to chance – mistakes made when the DNA was being copied from the parents' DNA, changes made by radiation, etc. Some changes will be damaging and make life more difficult for the animal unlucky enough to get that particular change. But others will be more fortunate, and will get a change that gives them some advantage in the struggle to survive. So, up to this point it is indeed all due to chance. But then comes the selection. Those animals that happen by chance to inherit a good change to their DNA (perhaps leading to sharper claws or faster running legs), are the ones more likely to survive and pass that good change on to future generations. Those that got the bad changes are the ones more likely to die out. So, in the end only the good changes get selected. It is this selection part of evolution by natural selection that is not purely down to chance.

Qu. 98

Could it be that some of the reason why we keep going to war is that there is a fighting code in our DNA? Do we tend naturally to be selfish because our ancestors developed that kind of streak in them in their struggle to grab short supplies of food and shelter? They survived because they made sure they got what was going, so today we should not be surprised to find that we tend to do the same.

That is the unpleasant side of genetically influenced behaviour. But a mother's or a father's natural tendency to look after and protect their children could be another.